Pediatric Reference Ranges

Fourth Edition

Pediatric Reference Ranges

Fourth Edition

Edited by

Steven J. Soldin, PhD
Department of Laboratory Medicine
Children's National Medical Center
and George Washington University
School of Medicine
and Georgetown Clinical Research Center
Georgetown University
Washington, DC

Carlo Brugnara, MD
Department of Laboratory Medicine
Children's Hospital and Harvard University
Boston, MA

Edward C. Wong, MD
Department of Laboratory Medicine
Children's National Medical Center
Washington, DC

AACC Press
2101 L Street, NW, Suite 202
Washington, DC 20037-1558

1 2 3 4 5 6 7 8 9 0 PCP 05 04 03

Printed in the United States of America

Library of Congress Cataloging-in-Publication Data

Pediatric reference ranges / edited by Steven J. Soldin, Carlo Brugnara, Edward C.
Wong—4th ed.
p. ; cm.
Includes bibliographical references.
ISBN 1-890883-88-3 (alk. paper)
1. Children—Diseases—Diagnosis. 2. Diagnosis, Laboratory. 3. Clinical chemistry. 4.
Reference values (Medicine) I. Soldin, Steven J., II. Brugnara, Carlo. III. Wong, Edward
C., date
[DNLM: 1. Chemistry, Clinical—Child. 2. Chemistry. Clinical—Infant. 3. Reference
Values. QY 16 P371 2003]
RJ51.L3P455 2003
618.92´0075—dc21 2003051810

We dedicate this book to our children and grandchildren.
We hope that children worldwide will benefit from this
fourth edition of Pediatric Reference Ranges.

CONTENTS

FOREWORD

Pediatric Reference Ranges is now in its fourth edition. This invaluable volume on pediatric laboratory medicine is a very important contribution to patient care, now made even more useful by the addition of a greatly expanded chemistry section and more complete hematology reference ranges. Every children's hospital, pediatric department, and clinical laboratory should have this well-designed and easy-to-read manual close at hand. The editors are to be commended for their splendid addition to the quality of laboratory diagnosis.

David G. Nathan, MD
Dana Farber Cancer Institute
Boston, Massachusetts

PREFACE

Children are unique individuals and not miniature adults. For example, neonates and premature infants have immature hepatic, renal, and pulmonary function. This affects the way these individuals handle drugs. Furthermore, their pediatric reference ranges for numerous analytes in many cases are clearly different from those found in older children and adults.

How Drug Metabolism Changes with Age

The majority of drugs are metabolized by the hepatic intestinal microsomal enzyme systems, which are under genetic control and subject to many factors that influence its activity. If, for instance, patients eat charbroiled meat or inhale the smoke of cigarettes, the activity of the microsomal system is enhanced, increasing the clearance of the drug being given and shortening its half life. Such patients may require dosage adjustments. Age is another major influence. In the neonate, the hepatic microsomal enzyme systems and renal and hepatic function are immature. Therefore, a much smaller dose per kilogram must be given to attain the same therapeutic concentration of a drug. During the first four months of life, infants need to be followed closely, because the enzyme activity is increasing, causing major changes in dosage requirement. In children from about six months of age to puberty, the hepatic microsomal enzyme system has approximately double the activity of the adult, requiring about double the dose per kilogram to achieve the same therapeutic concentration as an adult.

As the youth goes through puberty, the activity of the system begins decreasing, and eventually the individual has essentially the same hepatic microsomal enzyme system activity as an adult. One has to monitor closely if he/she is on drug therapy and going through puberty, because the activity of the enzyme system is changing a great deal over that time period.

The Concept of Reference Ranges

In order to determine if a patient suffers from a particular disease, clinicians will order tests that will confirm or refute their suspicions. A laboratory test is often the starting point for the differential diagnosis of a particular disease from among a spectrum of possible diseases. There are many different ways to establish age- and sex-related reference values, each method with its own inherent advantages and disadvantages. These include obtaining specimens from known "healthy" individuals and after removing the extreme outliers, tabulating values from the lowest to highest and removing the 0–2.5th percentile and the 97.5–100.00th percentile, thereby leaving a range of the 2.5–97.5th percentile for any particular analyte. One can also use the mean ± standard deviation (SD) approach, but for this method to be valid, the frequency distribution of values must be Gaussian (symmetrical or bell-shaped distribution). If the data are skewed and non-Gaussian, it can often be made Gaussian by plotting the log of the value instead of the value itself. Once a Gaussian distribution is obtained, one can calculate the reference range from the mean ± 2 SD.

Fig. 1 shows normal range plots for two different analytes. In (A) there is no overlap between the diseased and nondiseased populations. An example may be urinary vanillymandelic acid (VMA) or homovanillic acid (HVA) often ordered to confirm a diagnosis of pheochromocytoma or neuroblastoma. In (B), there is some overlap

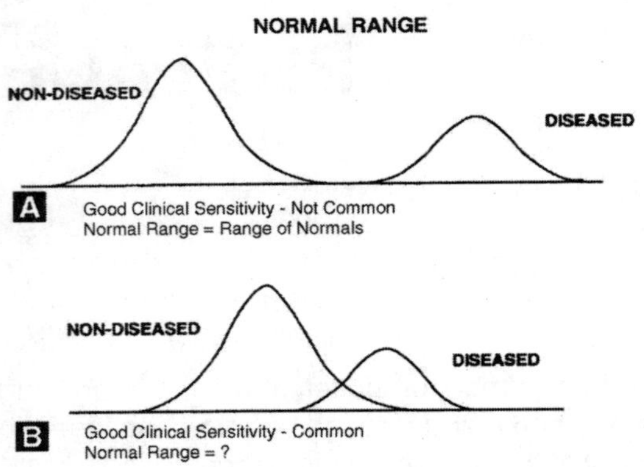

Fig. 1. Normal range plots for two analytes.

between the diseased and nondiseased populations. This situation is more common and choosing the normal reference range cut-off point here is somewhat arbitrary. In general, most clinicians are acquainted with a range that encompasses the 2.5–97.5th percentile of the population.

Hoffmann's Approach to Determining Reference Ranges

A third approach is to use the results obtained from hospitalized sick patients to develop the normal reference intervals. While this may seem to be a strange approach it has many advantages, especially in developing reference ranges for children. It is almost impossible to obtain a large enough population of healthy children between the ages of 1 day and 18 years who are prepared to donate blood specimens, or to obtain the informed consent necessary thereby allowing reference range studies to be performed (1). By using either Chauvenet's or Dixon's criteria for removing outliers and then plotting % cumulative frequency versus the laboratory value (or log of the value, if non-Gaussian distribution), one obtains a straight line, which deviates from linearity at both ends. This straight line can be extended to provide the 2.5th and 97.5th percentiles for the population being studied. This approach has been described in detail by Hoffmann (2), and an example for glucose is shown in Fig. 2. This approach, apart from its simplicity and appeal in pediatrics, has additional advantages for analytes such as T_4 (thyroxine) and cholesterol. T_4 values are significantly increased in a "sick" population. To assess the thyroid function in the "sick" population and differentiate those patients with thyroid disease from those with other ailments, it makes eminent sense to know the T_4 reference intervals for the "sick" population. Values above or below the 2.5th–97.5th range would then indicate the distinct possibility of thyroid disease. The same is true for cholesterol, values of which decrease somewhat in a "sick population."

Reference Ranges Are Method-Dependent

It is important to emphasize that reference ranges vary with the method (technology) used

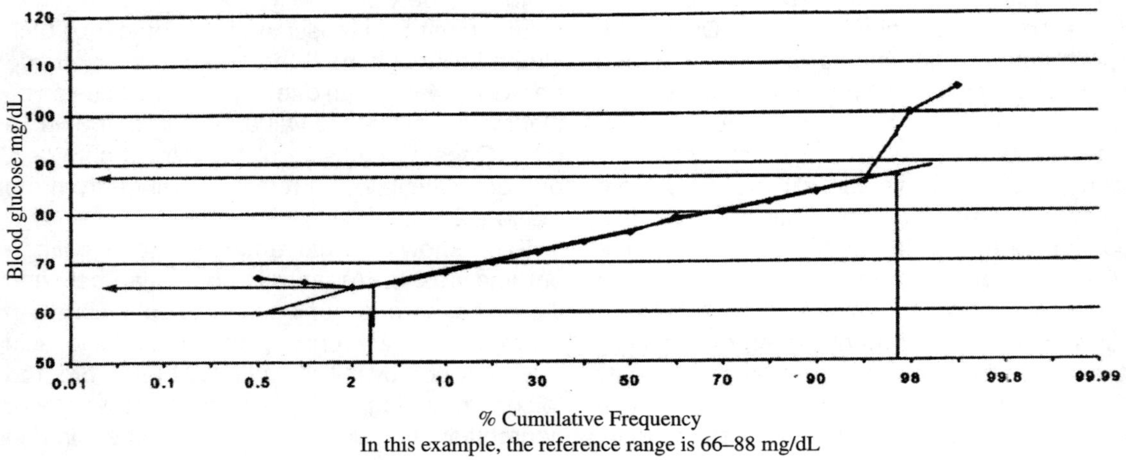

Fig. 2. Reference range for glucose.

to measure the analytes in question. In compiling the pediatric reference ranges for a textbook, I could not help noticing the very large differences one found for many endocrine tests, even if the same patient samples were used to establish these ranges. For example, when analyzing T_4, the male 1–30 day ranges using the Abbott IMx are 3.0–14.4 µg/dL (39–185 nmol/L), while the T_4 Gamma Coat (Baxter-Travenol Diagnostics, Cambridge, MA, USA) results are 5.9–21.5 µg/dL (76–276 nmol/L) (3). Such differences are extraordinary and indicate the considerable nonspecificity of antibodies used in immunoassays in the mid to late 1990's. One may well ask, what are we really measuring?

Children Are Not "Little Adults"

The reference ranges for many analytes vary significantly from birth through the end of puberty, as already stated. Neonates (and premature neonates), for example, have immature hepatic, renal, and pulmonary function. The normal values for blood gases, therefore, differ significantly from those of older children (pH 7.18–7.51 versus 7.35–7.44, respectively). The same is true for electrolytes such as sodium, potassium, and calcium, where reference ranges are somewhat broader than those found later in life. For example, for ionized calcium, the reference range for 0–1 month babies is 3.9–6.0 mg/dL (1.0–1.5 mmol/L), while in the adult it is 4.7–5.3 mg/dL (1.18–1.32 mmol/L).

Reference ranges for many enzymes vary significantly with age. For example, on the Vitros analyzer (Johnson & Johnson, Rochester, NY, USA) the male and female 97.5th percentile ranges for amylase are 30, 50, 80 and 100 U/L for 0–0.2, 0.2–0.5, 0.5–1.0, and 1–19 years, respectively. For aspartate aminotransferase (AST, SGOT), the male 2.5–97.5th percentiles using the Vitros analyzer are 30–100, 20–70, 15–45 U/L for 1–7 days, 8–30 days and 16–19 years, respectively.

Values for many endocrine tests undergo significant changes with onset of puberty. Male and female ranges can, and often do, differ significantly. In order to aid clinicians and pediatricians in interpreting laboratory data, we have recently gathered the literature available in this area, and published it in this textbook, *Pediatric Reference Ranges*. This book provides the reference ranges for a wide variety of analytes for children (males and females), from neonates to adolescents and young adults (3).

Conclusion

Anyone who uses reference ranges should realize that they are guidelines for the clinician and cannot be used as definitive indicators of health or pathologic states. Values for healthy and diseased individuals can overlap, and significant inter- and intra-individual differences exist even among healthy individuals (3).

References

1. Soldin SJ, Rifai N, Hicks JM, eds. Biochemical basis of pediatric disease, 3rd ed. Washington, DC: AACC Press, 1998.
2. Hoffmann RG. Statistics in the practice of medicine. JAMA 1963;185:864-73.
3. Soldin SJ, Brugnara SJ, Wong EC, eds. Pediatric reference ranges, 4th ed. Washington DC: AACC Press, 2003.

Steven J. Soldin, PhD, FACB, FCACB

INTRODUCTION

This book provides reference (normal) ranges on a wide variety of analytes for children from neonates to adolescents and young adults. We have put the emphasis on a user-friendly format. Each page has the same layout and provides information on the reference to the study. Whenever possible, the work is our own or from very recent literature. The type of fluid analyzed is given along with the methodology. We have tried to include at least two references for each analyte. We have quoted the different methodologies used. The names of the tests are reported alphabetically, and alternate names are sometimes given. In our studies, we have used a large number (n) of children in each age group, generally between 50 and 100. The reference values are reported by age and sex, and wherever possible we have included SI units so that this book can be used throughout the world. We have included the statistical basis by which the reference ranges were calculated and the population source.

We have purposely not burdened this book with clinical information that can be found elsewhere, especially in our previous book entitled *The Biochemical Basis of Pediatric Disease*, which Dr. Soldin co-edited with Dr. Jocelyn Hicks and Dr. Nader Rifai. It is our belief that, for the most part, when a physician or laboratory scientist seeks information about reference ranges, that is all he or she is interested in!

Anyone who uses reference ranges should realize that they are guidelines for the clinician and cannot be used as definitive indicators of health or pathological states. Values for healthy and diseased individuals often overlap, and, more importantly, considerable variations exist even among healthy individuals. Values also can vary according to the methodology used.

This book would not have been possible without the enormous help from American Medical Laboratories, Inc.® (now part of Quest Diagnostics) in Chantilly, Virginia. They provided the analyses for many of our studies as well as the complex statistical assistance that was necessary since our data was derived from patients of our hospitals.

Steven J. Soldin
Carlo Brugnara
Edward C. Wong

CHEMISTRY TESTS

ALANINE AMINOTRANSFERASE (ALT, SGPT)

		Male		Female	
Test	Age	n	U/L	n	U/L
1	1–7 d	109	6–40	84	7–40
	8–30 d	168	10–40	71	8–32
	1–3 mo	178	13–39	173	12–47
	4–6 mo	135	12–42	59	12–37
	7–12 mo	130	13–45	107	12–41
2	1–3 y	50*	5–45	50*	5–45
	4–6 y	40*	10–25	40*	10–25
	7–9 y	80*	10–35	80*	10–35
	10–11 y	27	10–35	34	10–30
	12–13 y	31	10–55	49	10–30
	14–15 y	26	10–45	52	5–30
	16–19 y	40	10–40	61	5–35
3	1–30 d	50	1–25	51	2–25
	31–365 d	91	4–35	76	3–30
	1–3 y	119	5–30	115	5–30
	4–6 y	114	5–20	101	5–25
	7–9 y	102	5–25	109	5–25
	10–18 y	280	5–30	269	5–20
4	1–7 d	**	20–54	**	21–54
	8–30 d		24–54		22–46
	1–3 mo		27–54		26–61
	4–6 mo		26–55		26–51
	7–12 mo		26–59		26–55
	1–3 y		19–59		24–59
	4–6 y		24–49		24–49
	10–11 y		24–49		24–44
	12–13 y		24–68		24–44
	14–15 y		24–59		19–44
	16–19 y		24–54		19–49

Specimen Type(s)	1–4	Plasma/serum
Reference(s)	1	Soldin SJ, Savwoir TV, Guo Y. Pediatric reference ranges for alkaline phosphatase, aspartate aminotransferase, and alanine aminotransferase in children less than 1 year old on the Vitros 500. Clin Chem 1997;43:S199. (Abstract)
	2	Lockitch G, Halstead AC, Albersheim S, et al. Age- and sex-specific pediatric reference intervals for biochemistry analytes as measured with the Ektachem-700 analyzer. Clin Chem 1988;34:1622–5.
	3	Soldin SJ, Bailey J, Bjorn S, et al. Pediatric reference ranges for ALT. Clin Chem 1995;41:S92–3. (Abstract)
	4	Ghoshal AK, Soldin SJ. Evaluation of the Dade Behring Dimension RxL: integrated chemistry system-pediatric reference ranges. Clin Chim Acta 2003;331:135–46.
Method(s)	1, 2	Vitros 500 (1) and 700 (2) (Ortho-Clinical Diagnostics, Raritan, NJ).
	3	Measured on the Hitachi 747 using Boehringer Mannheim reagents. (Boehringer Mannheim Diagnostics, Indianapolis, IN).
	4	Alanine aminotransferase (ALT) catalyzes the transamination of L-alanine to α-ketoglutarate (α-KG), forming L-glutamate and pyruvate. The pyruvate formed is reduced to lactate by lactate dehydrogenase (LDH) with simultaneous oxidation of reduced nicotinamide-adenine dinucleotide (NADH). Dade Behring Dimension RxL Analyzer (Dade Behring Inc., Newark, DE).
Comment(s)	1,3	Study used hospitalized patients and a computerized approach adapted from the Hoffmann technique to obtain the 2.5–97.5th percentiles.
	2	The study population was healthy children. Non-parametric methods were used to determine the 0.025 and 0.975 fractiles.
		*No significant differences were found for males and females. These ranges were therefore derived from combined data.
	4	**Reference ranges were obtained by comparing results from previously published data and using regression equations.

ALBUMIN

		Male			Female		
Test	**Age**	**n**	**g/dL**	**g/L**	**n**	**g/dL**	**g/L**
1	1–7 d	161	2.3–3.8	23–38	132	1.8–3.9	18–39
	8–30 d	252	2.0–4.5	20–45	124	1.8–4.4	18–44
	31–90 d	199	2.0–4.8	20–48	178	1.9–4.2	19–42
	91–180 d	135	2.1–4.9	21–49	121	2.2–4.4	22–44
	181 d–1 y	295	2.1–4.7	21–47	267	2.2–4.7	22–47
2	0–5 d (<2.5 kg)	30	2.0–3.6	20–36	30	2.0–3.6	20–36
	0–5 d (>2.5 kg)	93	2.6–3.6	26–36	93	2.6–3.6	26–36
	1–3 y	50	3.4–4.2	34–42	50	3.4–4.2	34–42
	4–6 y	38	3.5–5.2	35–52	38	3.5–5.2	35–52
	7–9 y	74	3.7–5.6	37–56	74	3.7–5.6	37–56
	10–19 y	332	3.7–5.6	37–56	332	3.7–5.6	37–56
3	1–30 d	73	2.6–4.1	26–41	51	2.7–4.3	27–43
	31–182 d	58	2.8–4.6	28–46	30	2.9–4.2	29–42
	183–365 d	29	2.8–4.8	28–48	42	3.3–4.8	33–48
	1–18 y	652	3.2–4.7	32–47	626	2.9–4.2	29–42
4	1–7 d	*	2.4–3.9	24–39	*	1.9–4.0	19–40
	8–30 d		2.1–4.5	21–45		1.9–4.4	19–44
	31–90 d		2.1–4.8	21–48		2.0–4.2	20–42
	91–180 d		2.2–4.9	22–49		2.3–4.4	23–44
	181 d–1 y		2.2–4.7	22–47		2.3–4.7	23–47
	1–3 y		3.5–4.2	35–42		3.5–4.7	35–47
	4–6 y		3.6–5.2	36–52		3.6–5.2	36–52
	7–9 y		3.8–5.6	38–56		3.8–5.6	38–56
	10–19 y		3.8–5.6	38–56		3.8–5.6	38–56

Specimen Type(s)	1–4	Plasma/serum
Reference(s)	1	Soldin SJ, Morse AS. Pediatric reference ranges for albumin and total protein in children <1 year old using the Vitros 500 analyzer. Clin Chem 1998;44:A15. (Abstract)
	2	Lockitch G, Halstead AC, Albersheim S, et al. Age- and sex-specific pediatric reference intervals for biochemistry analytes as measured with the Ektachem-700 analyzer. Clin Chem 1988;34:1622–5.
	3	Soldin SJ, Bjorn S, Beatey J, et al. Pediatric reference ranges for albumin, globulin and total protein on the Hitachi 747. Clin Chem 1995;41:S93. (Abstract)
	4	Ghoshal AK, Soldin SJ. Evaluation of the Dade Behring Dimension RxL: integrated chemistry system-pediatric reference ranges. Clin Chim Acta 2003;331:135–46.
Method(s)	1	Bromocresol green method. Vitros 500 (Ortho-Clinical Diagnostics, Raritan, NJ).
	2	Bromocresol green method. Vitros 700 (Ortho-Clinical Diagnostics, Raritan, NJ).
	3	Boehringer Mannheim albumin reagent (bromocresol green). Albumin was measured on the Hitachi 747 (Boehringer Mannheim Diagnostics, Indianapolis, IN).
	4	The albumin method is an adaptation of the bromocresol purple (BCP) dye-binding method. Dade Behring Dimension RxL Analyzer (Dade Behring Inc., Newark, DE).
Comment(s)	1,3	Study used hospitalized patients and a computerized approach adapted from the Hoffmann technique. Values are 2.5–97.5 percentiles.
	2	Healthy normal children. Values are 2.5–97.5 percentiles.
	4	*Reference ranges were obtained by comparing results from previously published data and using regression equations.

ALDOLASE

Male and Female		
Age	**n**	**U/L**
10–24 mo	40	3.4–11.8
25 mo–16 y	23	1.2–8.8
17–64 y	30	1.7–4.9

Specimen Type(s)	Serum
Reference(s)	Visnapuu LA, Karlson LK, Dubinsky EH, et al. Pediatric reference ranges for serum aldolase. Am J Clin Pathol 1989;91:476–7.
Method(s)	Aldolase Test Stat Pak (Behring Diagnostics, La Jolla, CA).
Comment(s)	Essentially healthy children and adults were used in the study. Results are mean ±2 SDs.

ALDOSTERONE

Male and Female				
Test	**Age**	**n**	**ng/dL**	**pmol/L**
1	6–9 y	25	1–24	28–666
	10–11 y	23	2–150	55–416
	12–14 y	27	1–22	28–610
	15–17 y	42	1–32	28–888
2	<1 y	*	5.8–110	160–3000
	1–<4 y		2.5–36	70–1000
	4–<10 y			
	P_1		1–22	30–600
	P_2		1.5–22	40–600
	P_3		1.5–22	40–600
	P_4		1.5–22	40–600
	P_5		1.5–22	40–600
3	Premature infants	**		
	26–28 w, day 4		5–635	139–17622
	31–35 w, day 4		19–141	527–3913
	Full-term infants			
	3 d		7–184	194–5106
	1 wk		5–175	139–4856
	1–12 mo		5–90	139–2498
	Children			
	1–2 y		7–54	194–1499
	2–10 y		3–35	83–971
	10–15 y		2–22	56–611

Specimen Type(s)	1,3	Serum
	2	Plasma/serum
Reference(s)	1	Pediatric endocrine testing. Nichols Institute, 1993:1.
	2	Soldin SJ, Rifai N, Hicks JM, eds. Biochemical basis of pediatric disease, 3rd ed. Washington, DC: AACC Press, 1988:233.
	3	Endocrinology expected values. Esoterix Endocrinology, Calabasas Hills, CA. ©2002 Esoterix, Inc. http://www.esoterix.com
Method(s)	1	Extraction, chromatography, radioimmunoassay.
	2	See reference.
	3	Not provided.
Comment(s)	1	Upright posture, normal sodium diet. Results are 2.5–97.5th percentiles.
	2	These reference ranges are for guidance only. Numbers not provided. P_1–P_5 refer to pubertal stages.
		*Numbers not provided.
	3	**Numbers not provided.
		Values are based on early morning samples from subjects on ad lib sodium intake. Diurnal variations and values in pediatric patients on different sodium diets are currently unavailable.

ALKALINE PHOSPHATASE (ALP)

		Male		Female	
Test	Age	n	U/L	n	U/L
1	1–7 d	141	77–265	109	65–270
	8–30 d	203	91–375	141	65–365
	1–3 mo	251	60–360	234	80–425
	4–6 mo	129	55–325	66	80–345
	7–12 mo	113	60–300	58	60–330
2*	1–3 y	50**	129–291	50**	129–291
	4–6 y	40**	134–346	40**	134–346
	7–9 y	80**	156–386	80**	156–386
	10–11 y	27	120–488	3	116–515
	12–13 y	31	178–455	49	93–386
	14–15 y	26	116–483	52	62–209
	16–19 y	40	58–237	61	45–116
3	1–30 d	60	75–316	75	48–406
	31–365 d	132	82–383	122	124–341
	1–3 y	136	104–345	111	108–317
	4–6 y	113	93–309	113	96–297
	7–9 y	124	86–315	104	69–325
	10–12 y	111	42–362	109	51–332
	13–15 y	126	74–390	105	50–162
	16–18 y	112	52–171	110	47–119
4	1–7 d	***	121–351	***	107–357
	8–30 d		138–486		107–474
	1–3 mo		101–467		125–547
	4–6 mo		94–425		125–449
	7–12 mo		101–394		101–431
	1–3 y		185–383		185–383
	4–6 y		191–450		191–450
	7–9 y		218–499		218–499
	10–11 y		174–624		169–657
	12–13 y		245–584		141–499
	14–15 y		169–618		103–283
	16–19 y		98–317		82–169

Specimen Type(s)	1,3,4	Plasma/serum
	2	Serum
Reference(s)	1	Soldin SJ, Savwoir TV, Guo Y. Pediatric reference ranges for alkaline phosphatase, aspartate aminotransferase, and alanine aminotransferase in children less than 1 year old on the Vitros 500. Clin Chem 1997;43:S199. (Abstract)
	2	Lockitch G, Halstead AC, Albersheim S, et al. Age- and sex-specific pediatric reference intervals for biochemistry analytes as measured with the Ektachem-700 analyzer. Clin Chem 1988;34:1622–5.
	3	Soldin SJ, Hicks JM, Bailey J, et al. Pediatric reference ranges for alkaline phosphatase on the Hitachi 747 analyzer. 1997;43:S198. (Abstract)
	4	Ghoshal AK, Soldin SJ. Evaluation of the Dade Behring Dimension RxL: integrated chemistry system-pediatric reference ranges. Clin Chim Acta 2003;331:135–46.
Method(s)	1,2	Vitros 500 and 700 (Ortho-Clinical Diagnostics, Raritan, NJ). p-Nitrophenylphosphate.
	3	Hitachi 747. Boehringer Mannheim reagents (p-Nitrophenylphosphate). (Boehringer Mannheim Diagnostics, Indianapolis, IN).
	4	Alkaline phosphatase catalyzes the transphosphorylation of p-nitrophenylphosphate (p-NPP) to p-nitrophenol (p-NP). Followed spectrophotometrically. Dade Behring Dimension RxL Analyzer (Dade Behring Inc., Newark, DE).
Comment(s)	1,3	Study used hospitalized patients and a computerized approach adapted from the Hoffmann technique to obtain the 2.5–97.5th percentiles.
	2	The study population was healthy children. Non-parametric methods were used to determine the 0.025 and 0.975 fractiles.
		*Due to changes made by the manufacturer in slide performance, these results are lower than those published by the author.
		**No significant differences were found for males and females. These ranges were therefore derived from combined data.
	4	***Reference ranges were obtained by comparing results from previously published data and using regression equations.

ALPHA$_1$-ANTITRYPSIN (α_1-AT)

		Male			Female		
Test	**Age**	**n**	**mg/dL**	**g/L**	**n**	**mg/dL**	**g/L**
1	0–5 d	73*	92–282	0.92–2.81	73*	92–282	0.92–2.81
	1–3 y	51*	94–156	0.94–1.56	51*	94–156	0.94–1.56
	4–6 y	39*	102–157	1.02–1.57	39*	102–157	1.02–1.57
	7–9 y	39*	102–157	1.02–1.57	39*	102–157	1.02–1.57
	10–13 y	36	104–159	1.04–1.59	45	106–171	1.06–1.71
	14–19 y	46	97–203	0.97–2.03	66	113–191	1.13–1.91

		Male and Female		
Test	**Age**	**n**	**mg/dL**	**g/L**
2	0–1 mo	60	79.4–222.7	0.79–2.23
	1–6 mo	45	71.0–190.1	0.71–1.90
	6 mo–2 y	82	60.1–160.6	0.60–1.61
	2–19 y	303	70.4–178.6	0.70–1.79

Specimen Type(s)

1,2 Serum

Reference(s)

The above values have been adjusted from those published to convert results to current IFCC units using the new IFCC guidelines/standards.

1 Lockitch G, Halstead AC, Quigley G, et al. Age- and sex-specific pediatric reference intervals: study design and methods illustrated by measurement of serum proteins with the Behring LN nephelometer. Clin Chem 1988;34:1618–21.

2 Davis ML, Austin C, Messmer BL, et al. IFCC-standardized pediatric reference intervals for 10 serum proteins using Beckman Array 360 system. Clin Biochem 1996;29:489–92.

Method(s)

1 Nephelometric with Behring LN nephelometer (Behring Diagnostics, Hoechst Canada, Inc., Montreal).

2 Beckman Array 360 (Beckman Instruments, Brea CA).

Comment(s)

1 Normal healthy children. This data set excluded children found to have SZ, MZ, or ZZ phenotypes by protease inhibitor typing.

Note: Type MM, normal; Type MS, normal variant (~8% of population); MZ, heterozygous for deficiency (<2% of population); ZZ, homozygous for deficiency. PI typing necessary if result is low. Because α_1-AT is an acute phase reactant, MZ individuals may have normal values. Values are 2.5–97.5th percentiles.

*Results from males and females combined. Values provided are 0.025–0.975 fractiles.

2 Samples were obtained from children attending outpatient clinics. Results are 2.5–97.5th percentiles.

ALPHA-FETOPROTEIN (AFP)

		Males		Non–Pregnant Females	
Test	**Age**	**n**	**ng/mL (µg/L)**	**n**	**ng/mL (µg/L)**
1	0–1 mo	71	0.6–16387	58	0.6–18964
	1–12 mo	113	0.6–28.3	102	0.6–77.0
	1–3 y	134	0.6–7.9	116	0.6–11.1
	4–6 y	101	0.6–5.6	118	0.6–4.2
	7–12 y	138	0.6–3.7	120	0.6–5.6
	13–18 y	145	0.6–3.9	122	0.6–4.2

			Male and Female
Test	**Age**	**n**	**ng/mL (µg/L)**
2	0–30 d	*	50.0–100,000
	1–3 mo		40.0–1000
	4 mo–18 y		0.0–12.0

Specimen Type(s)	1	Plasma
	2	Serum
Reference(s)	1	Soldin SJ, Hicks JM, Godwin ID, et al. Pediatric reference ranges for alpha-fetoprotein. Clin Chem 1992;38:959. (Abstract)
	2	Dugaw KA, Jack RM, Rutledge J. Pediatric reference ranges for ferritin and AFP on the Vitros ECi analyzer. Clin Chem 2001;47:A108–9. (Abstract)
Method(s)	1	Abbott EIA procedure (Abbott Laboratories, Abbott Park, IL).
	2	Chemiluminescent immunoassay, Vitros ECi (Ortho-Clinical Diagnostics, Raritan, NJ).
Comment(s)	1	Study used hospitalized patients and a computerized adaptation of the Hoffmann technique. Values are 2.5–97.5 percentiles.
	2	*Ages ranged from 1 h to 18 y with a total of 66 samples.

AMINO ACIDS (PLASMA)

Male and Female (nmol/mL)						
Test	Age	n	Alanine	β-Alanine	Anserine	α-Aminoadipic acid
1	Premature (first 6 wk)	*	212–504	0	—	0
	0–1 mo		131–710	0–10	0	0
	1–24 mo		143–439	0–7	0	0
	2–18 y		152–547	0–7	0	0
2	6 mo	*	182–396	—	—	—
	2 y		173–349	—	—	—
	6 y		182–319	—	—	—
	16 y		240–482	—	—	—

Specimen Type(s)	1,2	Plasma
Reference(s)	1	Shapira E, Blitzer MG, Miller JB, Africk DK, eds. Biochemical genetics: A laboratory manual. Oxford, UK: Oxford University Press, 1989;94–5.
	2	Lepage N, McDonald N, Dallaire L, et al. Age-specific distribution of plasma amino acid concentrations in a healthy pediatric population. Clin Chem 1997;43:2397–402.
Method(s)	1	Beckman amino acid analyzer. Beckman 6300 (Beckman Instruments Inc., Palo Alto, CA).
	2	Beckman amino acid analyzer. Beckman 7300 (Beckman Instruments Inc., Palo Alto, CA).
Comment(s)	1	Results are mean ±2 SDs.
	2	Results are 10–90th percentiles in healthy children.
	*See references for numbers.	

AMINO ACIDS (PLASMA)

Test	Age	n	α-Amino-n-butyric acid	λ-Amino-butyric acid	β-Amino-isobutyric acid	Arginine
			Male and Female (nmol/mL)			
1	Premature (first 6 wk)	*	14–52	0	0	34–96
	0–1 mo		8–24	0–2	0	6–140
	1–24 mo		3–26	0	0	12–133
	2–18 y		4–31	0	0	10–140
2	6 mo	*	—	—	—	43–120
	2 y		—	—	—	46–90
	6 y		—	—	—	50–99
	16 y		—	—	—	68–128

Specimen Type(s)	1,2	Plasma
Reference(s)	1	Shapira E, Blitzer MG, Miller JB, Africk DK, eds. Biochemical genetics: A laboratory manual. Oxford, UK: Oxford University Press, 1989:94–5.
	2	Lepage N, McDonald N, Dallaire L, et al. Age-specific distribution of plasma amino acid concentrations in a healthy pediatric population. Clin Chem 1997;43:2397–2402
Method(s)	1	Beckman amino acid analyzer. Beckman 6300 (Beckman Instruments Inc., Palo Alto, CA).
	2	Beckman amino acid analyzer. Beckman 7300 (Beckman Instruments Inc., Palo Alto, CA).
Comment(s)	1	Results are mean ±2 SDs.
	2	Results are 10–90th percentiles in healthy children.
	*See references for numbers.	

AMINO ACIDS (PLASMA)

Test	Age	n	Male and Female (nmol/mL)			
			Asparagine	Aspartic acid	Carnosine	Citrulline
1	Premature (first 6 wk)	*	90–295	24–50	—	20–87
	0–1 mo		29–132	20–129	0–19	10–45
	1–24 mo		21–95	0–23	0	3–35
	2–18 y		23–112	1–24	0	1–46
2	6 mo	*	31–56	4–18	—	14–32
	2 y		29–56	3–8	—	17–35
	6 y		31–67	3–6	—	23–37
	16 y		37–81	2–5	—	23–39

Specimen Type(s)	1,2	Plasma
Reference(s)	1	Shapira E, Blitzer MG, Miller JB, Africk DK, eds. Biochemical genetics: a laboratory manual. Oxford, UK: Oxford University Press, 1989:94–5.
	2	Lepage N, McDonald N, Dallaire L, et al. Age-specific distribution of plasma amino acid concentrations in a healthy pediatric population. Clin Chem 1997;43:2397–402.
Method(s)	1	Beckman amino acid analyzer. Beckman 6300 (Beckman Instruments Inc., Palo Alto, CA).
	2	Beckman amino acid analyzer. Beckman 7300 (Beckman Instruments Inc., Palo Alto, CA).
Comment(s)	1	Results are mean ±2 SDs.
	2	Results are 10–90th percentiles in healthy children.
	*See references for numbers.	

AMINO ACIDS (PLASMA)

Male and Female (nmol/mL)						
Test	**Age**	**n**	**Cystathionine**	**Cystine**	**Ethanolamine**	**Glutamic Acid**
1	Premature (first 6 wk)	*	5–10	15–70	—	107–276
	0–1 mo		0–3	17–98	0–115	62–620
	1–24 mo		0–5	16–84	0–4	10–133
	2–18 y		0–3	5–45	0–7	5–150
2	6 mo	*	—	21–53	—	31–113
	2 y		—	27–52	—	28–81
	6 y		—	33–54	—	13–65
	16 y		—	36–61	—	11–46

Specimen Type(s)	1,2	Plasma
Reference(s)	1	Shapira E, Blitzer MG, Miller JB, Africk DK, eds. Biochemical genetics: a laboratory manual. Oxford, UK: Oxford University Press, 1989:94–5.
	2	Lepage N, McDonald N, Dallaire L, et al. Age-specific distribution of plasma amino acid concentrations in a healthy pediatric population. Clin Chem 1997;43:2397–402.
Method(s)	1	Beckman amino acid analyzer. Beckman 6300 (Beckman Instruments Inc., Palo Alto, CA).
	2	Beckman amino acid analyzer. Beckman 7300 (Beckman Instruments Inc., Palo Alto, CA).
Comment(s)	1	Results are mean ±2 SDs.
	2	Results are 10–90th percentiles in healthy children.
	*See references for numbers.	

AMINO ACIDS (PLASMA)

			Male and Female (nmol/mL)			
Test	**Age**	**n**	**Glutamine**	**Glycine**	**Histidine**	**Homocystine**
1	Premature (first 6 wk)	*	248–850	298–602	72–134	3–20
	0–1 mo		376–709	232–740	30–138	0
	1–24 mo		246–1182	81–436	41–101	0
	2–18 y		254–823	127–341	41–125	0–5
2	6 mo	*	474–737	138–276	61–91	—
	2 y		473–692	138–276	61–91	—
	6 y		493–724	144–282	63–93	—
	16 y		551–797	183–322	77–107	—

Specimen Type(s)	1,2	Plasma
Reference(s)	1	Shapira E, Blitzer MG, Miller JB, Africk DK, eds. Biochemical genetics: A laboratory manual. Oxford, UK: Oxford University Press, 1989:94–5.
	2	Lepage N, McDonald N, Dallaire L, et al. Age-specific distribution of plasma amino acid concentrations in a healthy pediatric population. Clin Chem 1997;43:2397–402.
Method(s)	1	Beckman amino acid analyzer. Beckman 6300 (Beckman Instruments Inc., Palo Alto, CA).
	2	Beckman amino acid analyzer. Beckman 7300 (Beckman Instruments Inc., Palo Alto, CA).
Comment(s)	1	Results are mean ±2 SDs.
	2	Results are 10–90th percentiles in healthy children.
	*See references for numbers.	

AMINO ACIDS (PLASMA)

Male and Female (nmol/mL)						
Test	**Age**	**n**	**Hydroxylysine**	**Hydroxyproline**	**Isoleucine**	**Leucine**
1	Premature (first 6 wk)	*	0	0–80	23–85	151–220
	0–1 mo		0–7	0–91	26–91	48–160
	1–24 mo		0–7	0–63	31–86	47–155
	2–18 y		0–2	3–45	22–107	49–216
2	6 mo	*	—	—	39–76	77–153
	2 y		—	—	4–78	79–147
	6 y		—	—	40–69	86–136
	16 y		—	—	47–74	101–159

Specimen Type(s)	1,2	Plasma
Reference(s)	1	Shapira E, Blitzer MG, Miller JB, Africk DK, eds. Biochemical genetics: a laboratory manual. Oxford, UK: Oxford University Press, 1989:94–5.
	2	Lepage N, McDonald N, Dallaire L, et al. Age-specific distribution of plasma amino acid concentrations in a healthy pediatric population. Clin Chem 1997;43:2397–402.
Method(s)	1	Beckman amino cid analyzer. Beckman 6300 (Beckman Instruments Inc., Palo Alto, CA).
	2	Beckman amino acid analyzer. Beckman 7300 (Beckman Instruments Inc., Palo Alto, CA).
Comment(s)	1	Results are mean ±2 SDs.
	2	Results are 10–90th percentiles in healthy children.
	*See references for numbers.	

AMINO ACIDS (PLASMA)

Test	Age	n	Male and Female (nmol/mL)			
			Lysine	Methionine	1-Methyl-histidine	3-Methyl-histidine
1	Premature (first 6 wk)	*	128–255	37–91	4–28	5–33
	0–1 mo		92–325	10–60	0–43	0–5
	1–24 mo		52–196	9–42	0–44	0–5
	2–18 y		48–284	7–47	0–42	0–5
2	6 mo	*	87–171	14–38	—	—
	2 y		88–172	13–22	—	—
	6 y		96–181	14–25	—	—
	16 y		157–242	20–34	—	—

Specimen Type(s)	1,2	Plasma
Reference(s)	1	Shapira E, Blitzer MG, Miller JB, Africk DK, eds. Biochemical genetics: a laboratory manual. Oxford, UK: Oxford University Press, 1989:94–5.
	2	Lepage N, McDonald N, Dallaire L, et al. Age-specific distribution of plasma amino acid concentrations in a healthy pediatric population. Clin Chem 1997;43:2397–402.
Method(s)	1	Beckman amino acid analyzer. Beckman 6300 (Beckman Instruments Inc., Palo Alto, CA).
	2	Beckman amino acid analyzer. Beckman 7300 (Beckman Instruments Inc., Palo Alto, CA).
Comment(s)	1	Results are mean ±2 SDs.
	2	Results are 10–90th percentiles in healthy children.
	*See references for numbers.	

AMINO ACIDS (PLASMA)

Male and Female (nmol/mL)						
Test	Age	n	Ornithine	Phenylalanine	Phospho-ethanolamine	Phosphoserine
1	Premature (first 6 wk)	*	77–212	98–213	5–35	10–45
	0–1 mo		48–211	38–137	3–27	7–47
	1–24 mo		22–103	31–75	0–6	1–20
	2–18 y		10–163	26–91	0–69	1–30
2	6 mo	*	25–103	38–78	—	—
	2 y		24–60	39–65	—	—
	6 y		25–50	40–61	—	—
	16 y		37–62	47–74	—	—

Specimen Type(s)	1,2	Plasma
Reference(s)	1	Shapira E, Blitzer MG, Miller JB, Africk DK, eds. Biochemical genetics: a laboratory manual. Oxford, UK: Oxford University Press, 1989:94–5.
	2	Lepage N, McDonald N, Dallaire L, et al. Age-specific distribution of plasma amino acid concentrations in a healthy pediatric population. Clin Chem 1997;43:2397–402.
Method(s)	1	Beckman amino acid analyzer. Beckman 6300 (Beckman Instruments Inc., Palo Alto, CA).
	2	Beckman amino acid analyzer. Beckman 7300 (Beckman Instruments Inc., Palo Alto, CA).
Comment(s)	1	Results are mean ±2 SDs.
	2	Results are 10–90th percentiles in healthy children.
	*See references for numbers.	

AMINO ACIDS (PLASMA)

Test	Age	n	Proline	Sarcosine	Serine	Taurine
			Male and Female (nmol/mL)			
1	Premature (first 6 wk)	*	92–310	0	127–248	151–411
	0–1 mo		110–417	0–625	99–395	46–492
	1–24 mo		52–298	0	71–186	15–143
	2–18 y		59–369	0–9	69–187	10–170
2	6 mo	*	93–265	—	98–160	39–111
	2 y		93–220	—	97–154	39–80
	6 y		93–201	—	96–155	41–69
	16 y		113–271	—	101–177	41–66

Specimen Type(s)	1,2	Plasma
Reference(s)	1	Shapira E, Blitzer MG, Miller JB, Africk DK, eds. Biochemical genetics: a laboratory manual. Oxford, UK: Oxford University Press, 1989:94–5.
	2	Lepage N, McDonald N, Dallaire L, et al. Age-specific distribution of plasma amino acid concentrations in a healthy pediatric population. Clin Chem 1997;43:2397–402.
Method(s)	1	Beckman amino acid analyzer. Beckman 6300 (Beckman Instruments Inc., Palo Alto, CA).
	2	Beckman amino acid analyzer. Beckman 7300 (Beckman Instruments Inc., Palo Alto, CA).
Comment(s)	1	Results are mean ±2 SDs.
	2	Results are 10–90th percentiles in healthy children.
	*See references for numbers.	

AMINO ACIDS (PLASMA)

Test	Age	n	Threonine	Tryptophan	Tyrosine	Valine
			Male and Female (nmol/mL)			
1	Premature (first 6 wk)	*	150–330	28–136	147–420	99–220
	0–1 mo		90–329	0–60	55–147	86–190
	1–24 mo		24–174	23–71	22–108	64–294
	2–18 y		35–226	0–79	24–115	74–321
2	6 mo	*	61–162	34–73	43–108	135–260
	2 y		61–115	35–73	40–77	147–255
	6 y		65–125	37–76	39–65	165–234
	16 y		104–188	54–93	46–87	178–275

Specimen Type(s)	1,2	Plasma
Reference(s)	1	Shapira E, Blitzer MG, Miller JB, Africk DK, eds. Biochemical genetics: a laboratory manual. Oxford, UK: Oxford University Press, 1989:94–5.
	2	Lepage N, McDonald N, Dallaire L, et al. Age-specific distribution of plasma amino acid concentrations in a healthy pediatric population. Clin Chem 1997;43:2397–402.
Method(s)	1	Beckman amino acid analyzer. Beckman 6300 (Beckman Instruments Inc., Palo Alto, CA).
	2	Beckman amino acid analyzer. Beckman 7300 (Beckman Instruments Inc., Palo Alto, CA).
Comment(s)	1	Results are mean ±2 SDs.
	2	Results are 10–90th percentiles in healthy children.
	*See references for numbers.	

AMINO ACIDS (URINE)

Male and Female (nmol/mg Creatinine)					
Age	n	Alanine	β-Alanine	Anserine	α-Aminoadipic acid
Premature (first 6 wk)	*	1320–4040	1020–3500	—	70–460
0–1 mo		982–3055	25–288	0–3	0–180
1–24 mo		767–6090	0–297	0–5	45–268
2–18 y		231–915	0–65	0	2–88

Specimen Type(s)	Urine
Reference(s)	Shapira E, Blitzer MG, Miller JB, Africk DK, eds. Biochemical genetics: a laboratory manual. Oxford, UK: Oxford University Press, 1989:96–7.
Method(s)	Beckman amino acid analyzer. Beckman 6300 (Beckman Instruments Inc., Palo Alto, CA).
Comment(s)	Results are mean ±2 SDs. *See reference for numbers.

AMINO ACIDS (URINE)

Male and Female (nmol/mg Creatinine)					
Age	n	α-Amino-n-butyric acid	γ-Amino-butyric acid	β-Amino-isobutyric acid	Arginine
Premature (first 6 wk)	*	50–710	20–260	50–470	190–820
0–1 mo		8–65	0–15	421–3133	35–214
1–24 mo		30–136	0–105	802–4160	38–165
2–18 y		0–77	15–30	291–1482	31–109

Specimen Type(s)	Urine
Reference(s)	Shapira E, Blitzer MG, Miller JB, Africk DK, eds. Biochemical genetics: a laboratory manual. Oxford, UK: Oxford University Press, 1989:96–7.
Method(s)	Beckman amino acid analyzer. Beckman 6300 (Beckman Instruments Inc., Palo Alto, CA).
Comment(s)	Results are mean ±2 SDs. *See reference for numbers.

AMINO ACIDS (URINE)

Age	n	Asparagine	Aspartic acid	Carnosine	Citrulline
Male and Female (nmol/mg Creatinine)					
Premature (first 6 wk)	*	1350–5250	580–1520	260–370	240–1320
0–1 mo		185–1550	336–810	97–665	27–181
1–24 mo		252–1280	230–685	203–635	22–180
2–18 y		72–332	0–120	72–402	10–99

Specimen Type(s)	Urine
Reference(s)	Shapira E, Blitzer MG, Miller JB, Africk DK, eds. Biochemical genetics: a laboratory manual. Oxford, UK: Oxford University Press, 1989:96–7.
Method(s)	Beckman amino acid analyzer. Beckman 6300 (Beckman Instruments Inc., Palo Alto, CA).
Comment(s)	Results are mean ±2 SDs. *See reference for numbers.

AMINO ACIDS (URINE)

Age	n	Cystathionine	Cystine	Ethanolamine	Glutamic Acid
Male and Female (nmol/mg Creatinine)					
Premature (first 6 wk)	*	260–1160	480–1690	—	380–3760
0–1 mo		16–147	212–668	840–3400	70–1058
1–24 mo		33–470	68–710	0–2230	54–590
2–18 y		0–26	25–125	0–530	0–176

Specimen Type(s)	Urine
Reference(s)	Shapira E, Blitzer MG, Miller JB, Africk DK, eds. Biochemical genetics: a laboratory manual. Oxford, UK: Oxford University Press, 1989:96–7.
Method(s)	Beckman amino acid analyzer. Beckman 6300 (Beckman Instruments Inc., Palo Alto, CA).
Comment(s)	Results are mean ±2 SDs. *See reference for numbers.

AMINO ACIDS (URINE)

Age	n	Glutamine	Glycine	Histidine	Homocystine
Male and Female (nmol/mg Creatinine)					
Premature (first 6 wk)	*	520–1700	7840–23600	1240–7240	580–2230
0–1 mo		393–1042	5749–16423	908–2528	0–88
1–24 mo		670–1562	3023–11148	815–7090	6–67
2–18 y		369–1014	897–4500	644–2430	0–32

Specimen Type(s)	Urine
Reference(s)	Shapira E, Blitzer MG, Miller JB, Africk DK, eds. Biochemical genetics: a laboratory manual. Oxford, UK: Oxford University Press, 1989:96–7.
Method(s)	Beckman amino acid analyzer. Beckman 6300 (Beckman Instruments Inc., Palo Alto, CA).
Comment(s)	Results are mean ±2 SDs. *See reference for numbers.

AMINO ACIDS (URINE)

Age	n	Hydroxylysine	Hydroxyproline	Isoleucine	Leucine
Male and Female (nmol/mg Creatinine)					
Premature (first 6 wk)	*	—	560–5640	250–640	190–790
0–1 mo		10–125	40–440	125–390	78–195
1–24 mo		0–97	0–4010	38–342	70–570
2–18 y		40–102	0–3300	10–126	30–500

Specimen Type(s)	Urine
Reference(s)	Shapira E, Blitzer MG, Miller JB, Africk DK, eds. Biochemical genetics: a laboratory manual. Oxford, UK: Oxford University Press, 1989:96–7.
Method(s)	Beckman amino acid analyzer. Beckman 6300 (Beckman Instruments Inc., Palo Alto, CA).
Comment(s)	Results are mean ±2 SDs. *See reference for numbers.

AMINO ACIDS (URINE)

Age	n	Lysine	Methionine	1-Methyl-histidine	3-Methyl-histidine
Male and Female (nmol/mg Creatinine)					
Premature (first 6 wk)	*	1860–15460	500–1230	170–880	420–1340
0–1 mo		270–1850	342–880	96–499	189–680
1–24 mo		189–850	174–1090	106–1275	147–391
2–18 y		153–634	16–114	170–1688	182–365

Specimen Type(s)	Urine
Reference(s)	Shapira E, Blitzer MG, Miller JB, Africk DK, eds. Biochemical genetics: a laboratory manual. Oxford, UK: Oxford University Press, 1989:96–7.
Method(s)	Beckman amino acid analyzer. Beckman 6300 (Beckman Instruments Inc., Palo Alto, CA).
Comment(s)	Results are mean ±2 SDs. *See reference for numbers.

AMINO ACIDS (URINE)

Age	n	Ornithine	Phenylalanine	Phospho-ethanolamine	Phosphoserine
Male and Female (nmol/mg Creatinine)					
Premature (first 6 wk)	*	260–3350	920–2280	80–340	500–1690
0–1 mo		118–554	91–457	0–155	150–339
1–24 mo		55–364	175–1340	108–533	112–304
2–18 y		31 –91	61–314	18–150	70–138

Specimen Type(s)	Urine
Reference(s)	Shapira E, Blitzer MG, Miller JB, Africk DK, eds. Biochemical genetics: a laboratory manual. Oxford, UK: Oxford University Press, 1989:96–7.
Method(s)	Beckman amino acid analyzer. Beckman 6300 (Beckman Instruments Inc., Palo Alto, CA).
Comment(s)	Results are mean ±2 SDs. *See reference for numbers.

AMINO ACIDS (URINE)

Male and Female (nmol/mg Creatinine)					
Age	n	Proline	Sarcosine	Serine	Taurine
Premature (first 6 wk)	*	1350–10460	0	1680–6000	5190–23620
0–1 mo		370–2323	0–56	1444–3661	1650–6220
1–24 mo		254–2195	30–358	845–3190	545–3790
2–18 y		0	0–26	362–1100	639–1866

Specimen Type(s)	Urine
Reference(s)	Shapira E, Blitzer MG, Miller JB, Africk DK, eds. Biochemical genetics: a laboratory manual. Oxford, UK: Oxford University Press, 1989:96–7.
Method(s)	Beckman amino acid analyzer. Beckman 6300 (Beckman Instruments Inc., Palo Alto, CA).
Comment(s)	Results are mean ±2 SDs. *See reference for numbers.

AMINO ACIDS (URINE)

Male and Female (nmol/mg Creatinine)					
Age	n	Threonine	Tryptophan	Tyrosine	Valine
Premature (first 6 wk)	*	840–5700	0	1090–6780	180–890
0–1 mo		445–1122	0	220–1650	113–369
1–24 mo		252–1528	0–93	333–1550	99–316
2–18 y		121–389	0–108	122–517	58–143

Specimen Type(s)	Urine
Reference(s)	Shapira E, Blitzer MG, Miller JB, Africk DK, eds. Biochemical genetics: a laboratory manual. Oxford, UK: Oxford University Press, 1989:96–7.
Method(s)	Beckman amino acid analyzer. Beckman 6300 (Beckman Instruments Inc., Palo Alto, CA).
Comment(s)	Results are mean ±2 SDs. *See reference for numbers.

AMMONIA

Male and Female			
Test	**Age**	**n**	**µmol/L**
1	All ages	*	<33
2	Newborn	**	<50
	Child and adult		<35
3	<30 d	87	21–95
	1–12 mo	91	18–74
	1–14 y	94	17–68
	<14 y	182	22–66
	Men	89	21–71
	Women	93	19–63
4	All ages	***	<51

Specimen Type(s)	1,2	Plasma
	3	Whole blood, potassium EDTA as anticoagulant
	4	Plasma/serum
Reference(s)	1	Children's National Medical Center. Unpublished data.
	2	Reference values and SI unit information. The Hospital for Sick Children, Toronto, 1993.
	3	Diaz J, Tornel PL, Martinez P. Reference intervals for blood ammonia in healthy subjects, determined by microdiffusion. Clin Chem 1995;41:1048. (Letter)
	4	Ghoshal AK, Soldin SJ. Evaluation of the Dade Behring Dimension RxL: integrated chemistry system-pediatric reference ranges. Clin Chim Acta 2003;331:135–46.
Method(s)	1,2	Vitros 500 and 700 (Ortho-Clinical Diagnostics, Raritan, NJ).
	3	Ammonia Checker II (Menarini Diagnostics, Florence, Italy).
	4	Glutamate dehydrogenase (GLDH) catalyzes the condensation of ammonia and α-ketoglutarate (α-KG) with simultaneous oxidation of reduced nicotinamide adenine dinucleotide phosphate (NADPH). Dade Behring Dimension RxL Analyzer (Dade Behring Inc., Newark, DE).
Comment(s)	1	Values in premature infants may go as high as 50 µmol/L. Values are 2.5–97.5th percentiles.
		*Numbers not given.
	2	**Numbers not given.
	3	Values provided are 5–95th percentiles.
	4	***Reference ranges were obtained by comparing results from previously published data and using regression equations.

AMYLASE

Male and Female			
Test	**Age**	**n**	**U/L**
1	1–30 d	76	0–6
	31–182 d	110	1–17
	183–365 d	54	6–44
	1–3 y	148	8–79
	4–9 y	96	16–91
	10–18 y	142	19–76
2	0–0.2 y	55	<30
	0.2–0.5 y	81	<50
	0.5–1.0 y	170	<80
3	1– 19 y	470	30–100
4	1–30 d	*	<18
	31–182 d		<43
	183–365 d		<81
	1–3 y		<106
	4–9 y		<106
	10–18 y		<106

Specimen Type(s)	1,2	Plasma
	3	Serum
	4	Plasma/serum
Reference(s)	1	Soldin SJ, Hicks JM, Bailey J, et al. Pediatric reference ranges for amylase. Clin Chem 1995;41:S94. (Abstract)
	2	Soldin SJ, Rakotoarisoa FTS. Pediatric reference ranges for amylase and cholesterol on the Kodak Ektachem 500 in the first year of life. Clin Chem 1996;42:S308. (Abstract)
	3	Lockitch G, Halstead AC, Albersheim S, et al. Age- and sex-specific pediatric reference intervals for biochemistry analytes as measured with the Ektachem-700 analyzer. Clin Chem 1988;34:1622–5.
	4	Ghoshal AK, Soldin SJ. Evaluation of the Dade Behring Dimension RxL: integrated chemistry system-pediatric reference ranges. Clin Chim Acta 2003;331:135–46.
Method(s)	1	Hitachi 717 using Boehringer Mannheim reagents (Boehringer Mannheim Diagnostics, Indianapolis, IN).
	2,3	Amylopectin. Vitros 500 and 700 (Ortho-Clinical Diagnostics, Raritan, NJ).
	4	The amylase method on the Dimension® system uses a chromogenic substrate, 2-chloro-4-nitrophenol linked with maltotriose. The direct reaction of α-amylase with the substrate results in the formation of 2-chloro-4 nitrophenol, which is monitored spectrophotometrically. Dade Behring Dimension RxL Analyzer (Dade Behring Inc., Newark, DE).
Comment(s)	1	Study used hospitalized patients and a computerized approach to removing outliers. Values are 2.5–97.5th percentiles.
	2	Study used hospitalized patients and a computerized approach to removing outliers. Values are 97.5th percentiles.
	3	The study population was healthy children. Non-parametric methods were used to determine the 0.025 and 0.975 fractiles.
	4	*Reference ranges were obtained by comparing results from previously published data and using regression equations.

ANDROSTENEDIONE

		Male			Female		
Test	**Age**	**n**	**ng/dL**	**nmol/L**	**n**	**ng/dL**	**nmol/L**
1	1–5 mo	9	5–45	0.2–1.6	5	5–35	0.2–1.2
	6–11 mo	8	5–30	0.2–1.0	6	5–25	0.2–0.9
	1–5 y	31	5–45	0.2–1.6	18	5–40	0.2–1.4
	6–9 y	17	5–55	0.2–1.9	9	5–45	0.2–1.6
	10–11 y	7	10–30	0.3–1.0	11	25–80	0.9–2.8
	12–14 y	31	20–85	0.7–3.0	23	15–175	0.5–6.1
	15–17 y	16	35–100	1.2–3.5	10	55–200	1.9–7.0
	Adults	*	50–250	1.7–8.7	*	50–250	1.7–8.7
2	0–1 d	**	15–145	0.5–5.0	**	15–175	0.5–6.0
	1–7 d		20–110	0.7–3.8		25–95	0.9–3.3
	7–28 d		25–160	0.9–5.5		9–90	0.3–3.0
	1–12 mo		6–90	0.2–3.0		6–145	0.2–5.0
	1–<4 y		6–35	0.2–1.2		6–45	0.2–1.5
	4–<10 y						
	P_1		25–190	0.8–3.0		3–60	0.1–2.0
	P_2		15–120	0.5–4.0		30–145	1.0–5.0
	P_3		18–145	0.6–5.0		30–200	1.0–7.0
	P_4		15–220	0.5–7.5		18–260	0.6–9.0
	P_5		40–260	1.3–9.0		18–260	0.6–9.0

Test	Male and Female			
	Age	n	ng/dL	nmol/L
3	Premature infants	***		
	26–28 w, day 4		92–282	3.2–9.8
	31–35 w, day 4		80–446	2.8–15.5
	Full-term infants			
	1–7 d[a]		20–290[b]	0.7–10.1
	1–12 mo[c]		6–68	0.2–2.4
	Prepubertal children			
	1–10 y		8–50	0.3–1.7

Test	Male			Female[d]		
	Age	ng/dL	nmol/L	Age	ng/dL	nmol/L
3	Puberty			Puberty		
	Tanner Stage[e]			Tanner Stage[e]		
	1 (<9.8 y)	8–50	0.3–1.7	1 (<9.2 y)	8–50	0.3–1.7
	2 (9.8–14.5 y)	31–65	1.1–2.3	2 (9.2–13.7 y)	42–100	1.5–3.5
	3 (10.7–15.4 y)	50–100	1.7–3.5	3 (10.0–14.4 y)	80–190	2.8–6.6
	4 (11.8–16.2 y)	48–140	1.7–4.9	4 (10.7–15.6 y)	77–225	2.7–7.8
	5 (12.8–17.3 y)	65–210	2.3–7.3	5 (11.8–18.6 y)	80–240	2.8–8.4
	n***			**n*****		

Specimen Type(s)	1,3	Serum
	2	Plasma/serum
Reference(s)	1	Pediatric endocrine testing. Nichols Institute, 1993:5.
	2	Soldin SJ, Rifai N, Hicks JM, eds. Biochemical basis of pediatric disease, 3rd ed. Washington, DC: AACC Press, 1998:233.
	3	Endocrinology expected values. Esoterix Endocrinology, Calabasas Hills, CA. ©2002 Esoterix, Inc. http://www.esoterix.com
		[a]Forest MG, Bertrand J. Sexual steroids in neonatal period. Abstracts of the 7th Congress of the International Study Group for Steroids. J Steroid Biochem 1975;6:xxiv.
		[e]Tanner JM, Whitehouse RH. Clinical longitudinal standards for height, weight, height velocity, and the stages of puberty. Arch Dis Childhood 1976;51:170.
Method(s)	1	Extraction, chromatography, radioimmunoassay.
	2	See reference.
	3	Not provided.
Comment(s)	1	*Numbers not provided.
		Results are 2.5–97.5th percentiles.
	2	**Numbers not provided.
		P_1–P_5 refers to pubertal stages. These ranges are for guidance only.
	3	***Numbers not provided.
		[b]Levels decrease rapidly to a range of 18–80 ng/dL after one week.
		[c]Androstenedione gradually decreases during the first six months to prepubertal levels.
		[d]Day of cycle not determined for prepubertal females.
		[e]As used herein, Tanner stages in males encompass development of both pubic hair and genitalia. In females, each stage encompasses both pubic hair and breast development. While this expands the chronological age range for each stage of pubertal development, it results in a better correlation with hormonal values.

APOLIPOPROTEIN A-1

		Male			Female		
Test	**Age**	**n**	**g/L**	**mg/dL**	**n**	**g/L**	**mg/dL**
1	5 y	*	1.04–1.67	104–167	*	1.08–1.60	108–160
	10 y		1.05–1.60	105–160		1.04–1.57	104–157
	15 y		1.02–1.57	102–157		1.04–1.56	104–156
	20 y		1.04–1.65	104–165		1.08–1.65	108–165
2	4–5 y	**	1.09–1.72	109–172	**	1.04–1.63	104–163
	6–11 y		1.11–1.77	111–177		1.10–1.66	110–166
	12–19 y		0.99–1.65	99–165		1.05–1.80	105–180

		Male and Female		
Test	**Age**	**n**	**g/L**	**mg/dL**
3	2–12 mo	100	0.79–1.87	79–187
	2–10 y	120	1.07–1.79	107–179

Specimen Type(s)	1	Plasma
	2,3	Serum
Reference(s)	1	Kottke BA, Moll PP, Michels VV, et al. Levels of lipids, lipoproteins, and apolipoproteins in a defined population. Mayo Clin Proc 1991;66:1198–208.
	2	Bachorik PS, Lovejoy KL, Carroll MD, et al. Apolipoprotein B and A1 distributions in the United States, 1988–1991: results of the National Health and Nutrition Examination Survey III (NHANES III). Clin Chem 1997;43:2364–78
	3	Baroni S, Scribano D, Valentini P, et al. Serum apolipoprotein A1, B, CII, CIII, E, and lipoprotein (a) levels in children. Clin Biochem 1996;29:603–5.
Method(s)	1	Radioimmunoassay. See reference.
	2	Radial immunodiffusion and rate immunonephelometry. See reference above.
	3	Behring nephelometry using Behring reagents (Behring Diagnostics, Westwood, MA).
Comment(s)	1	Results are 5–95th percentiles read off graphs in above reference. *See references for numbers.
	2	Healthy children were studied. The results are the 5–95th percentiles. **The numbers studied are not provided.
	3	Results are the 2.5–97.5th percentiles for healthy children.

APOLIPOPROTEIN (a)

Male and Female			
Test	Age	n	U/L
1	2 d	29	0–59
	3 d	151	0–57
	4 d	459	0–69
	5 d	341	0–66
	6 d	30	0–78
	7 d	22	0–71
Test	Age	n	mg/L
2	2–12 mo	100	0–114
	2–10 y	120	0–178

Specimen Type(s)	1	Dried blood spot
	2	Serum
Reference(s)	1	Wang XL, Wilcken DEL, Dudman NPB. Neonatal apo A-1, apo-B and apo(a) levels in dried blood spots in an Australian population. Ped Res 1990;28:496–501.
	2	Baroni S, Scribano D, Valentini P, et al. Serum apolipoprotein A1, B, CII, CIII, E and lipoprotein (a) levels in children. Clin Biochem 1996;29:603–5.
Method(s)	1	Modification of RIA serum assay for apo(a). (Pharmacia Diagnostics AB, Uppsala, Sweden).
	2	EIA, Macra Lpa test kit (Strategic Diagnostics, Newark, DE).
Comment(s)	1	Healthy neonates were studied. Results are mean ±2 SDs.
	2	Results are the 2.5–97.5th percentiles for healthy children.

APOLIPOPROTEIN B

		Male			Female		
Test	Age	n	g/L	mg/dL	n	g/L	mg/dL
1	5 y	*	0.51–0.80	51–80	*	0.54–0.87	54–87
	10 y		0.50–0.88	50–88		0.54–0.87	54–87
	15 y		0.48–0.85	48–85		0.54–0.87	54–87
	20 y		0.48–0.90	48–90		0.54–0.85	54–85
2	4–5 y	**	0.58–1.03	58–103	**	0.58–1.04	58–104
	6–11 y		0.56–1.05	56–105		0.57–1.13	57–113
	12–19 y		0.55–1.10	55–110		0.53–1.19	53–119

		Male and Female		
Test	Age	n	mg/dL	g/L
3	2–12 mo	100	41–105	0.41–1.05
	2–10 y	120	44–112	0.44–1.12

Specimen Type(s)	1	Plasma
	2,3	Serum
Reference(s)	1	Kottke BA, Moll PP, Michels V, et al. Levels of lipids, lipoproteins, and apolipoproteins in a defined population. Mayo Clin Proc 1991;66:1198–208.
	2	Bachorik PS, Lovejoy KL, Carroll MD, et al. Apolipoprotein B and A1 distributions in the United States, 1988–1991: results of the National Health and Nutrition Examination Survey III (NHANES III). Clin Chem 1997;43:2364–78.
	3	Baroni S, Scribano D, Valentini P, et al. Serum apolipoprotein A1, B, CII, CIII, E and lipoprotein (a) levels in children. Clin Biochem 1996;29:603–5.
Method(s)	1	Enzyme linked immunosorbent assay. See reference.
	2	Radial immunodiffusion and rate nephelometry. See reference above.
	3	Behring nephelometer using Behring reagents (Behring Diagnostics, Westwood, MA).
Comment(s)	1	Results are 5–95th percentiles read off graphs in above reference.
		*See reference for numbers.
	2	Healthy children were studied. The results are the 5–95th percentiles.
		**The numbers studied are not provided.
	3	Results are the 2.5th and 97.5th percentiles for healthy children.

APOLIPOPROTEIN CII

Male and Female		
Age	n	mg/L
2–12 mo	100	15–79
2–10 y	120	9–73

Specimen Type(s)	Serum
Reference(s)	Baroni S, Scribano D, Valentini P, et al. Serum apolipoprotein A1, B, CII, CIII, E, and lipoprotein (a) levels in children. Clin Biochem 1996;29:603–5.
Method(s)	Turbidimetric immunoassay (Turbilinear "Eiken" Poli, Milan, Italy).
Comment(s)	Results are the 2.5–97.5th percentiles for healthy children.

APOLIPOPROTEIN CIII

Male and Female		
Age	n	mg/L
2–12 mo	100	18–134
2–10 y	120	25–113

Specimen Type(s)	Serum
Reference(s)	Baroni S, Scribano D, Valentini P, et al. Serum apolipoprotein A1, B, CII, CIII, E, and lipoprotein (a) levels in children. Clin Biochem 1996;29:603–5.
Method(s)	Turbidimetric immunoassay (Turbilinear "Eiken" Poli, Milan, Italy).
Comment(s)	Results are the 2.5–97.5th percentiles for healthy children.

APOLIPOPROTEIN E

Male and Female		
Age	n	mg/L
2–12 mo	100	23–59
2–10 y	120	19–59

Specimen Type(s)	Serum
Reference(s)	Baroni S, Scribano D, Valentini P, et al. Serum apolipoprotein A1, B, CII, CIII, E, and lipoprotein (a) levels in children. Clin Biochem 1996;29:603–5.
Method(s)	Turbidimetric immunoassay (Turbilinear "Eiken" Poli, Milan, Italy).
Comment(s)	Results are the 2.5–97.5th percentiles for healthy children.

ASPARTATE AMINOTRANSFERASE (AST, SGOT)

		Male		Female	
Test	Age	n	U/L	n	U/L
1	1–7 d	69	30–100	52	24–95
	8–30 d	148	20–70	84	24–72
	1–3 mo	160	22–63	131	20–64
	4–6 mo	133	13–65	83	20–63
	7–12 mo	131	25–55	142	22–63
2	1–3 y	50*	20–60	50*	20–60
	5–6 y	40*	15–50	40*	15–50
	7–9 y	80*	15–40	80*	15–40
	10–11 y	27	10–60	34	10–40
	12–13 y	31	15–40	49	10–30
	14–15 y	26	15–40	52	10–30
	16–19 y	40	15–45	61	5–30
3	1–30 d	74	<51	57	<49
	31–365 d	83	<65	71	<79
	1–3 y	134	<56	108	<69
	4–6 y	85	<48	84	<59
	7–9 y	122	<42	96	<41
	10–12 y	104	<38	62	<37
	13–15 y	88	<39	86	<32
	16–18 y	62	<39	78	<30
4	1–7 d	**	26–98	**	20–93
	8–30 d		16–67		20–69
	1–3 mo		16–60		16–61
	4–6 mo		16–62		16–60
	7–12 mo		16–52		16–60
	1–3 y		16–57		16–57
	5–6 y		10–47		10–47
	7–9 y		10–36		5–36
	12–15 y		10–36		5–26
	16–19 y		10–41		0–26

Specimen Type(s)	1,3,4	Plasma/serum
	2	Plasma
Reference(s)	1	Soldin SJ, Savwoir TV, Guo Y. Pediatric reference ranges for alkaline phosphatase, aspartate aminotransferase, and alanine aminotransferase in children less than 1 year old on the Vitros 500. Clin Chem 1997;43:S199. (Abstract)
	2	Lockitch G, Halstead AC, Albersheim S, et al. Age- and sex-specific pediatric reference intervals for biochemistry analytes as measured with the Ektachem-700 analyzer. Clin Chem 1988;34:1622–5.
	3	Soldin SJ, Hicks JM, Bailey J, et al. Pediatric reference ranges for AST. Clin Chem 1995;44:S94. (Abstract)
	4	Ghoshal AK, Soldin SJ. Evaluation of the Dade Behring Dimension RxL: integrated chemistry system-pediatric reference ranges. Clin Chim Acta 2003;331:135–46.
Method(s)	1,2	Vitros 500 (1) and 700 (2) (Ortho-Clinical Diagnostics, Raritan, NJ).
	3	Measured on the Hitachi 747 using Boehringer Mannheim reagents (Boehringer Mannheim Diagnostics, Indianapolis, IN).
	4	Aspartate aminotransferase catalyzes the transamination from L-aspartate to α-ketoglutarate, forming L-glutamate and oxalacetate. The oxalacetate formed is reduced to malate by malate dehydrogenase (MDH) with simultaneous oxidation of reduced nicotinamide adenine dinucleotide (NADH). Dade Behring Dimension RxL Analyzer (Dade Behring Inc., Newark, DE).
Comment(s)	1,3	Study used hospitalized patients and a computerized approach adapted from the Hoffmann technique to obtain the 2.5–97.5th percentiles (Test 1) and 97.5th percentile (Test 3).
	2	The study population was healthy children. Non-parametric methods were used to determine the 0.025 and 0.975 fractiles.
		*No significant differences were found for males and females. These ranges were therefore derived from combined data.
	4	**Reference ranges were obtained by comparing results from previously published data and using regression equations.

BASE EXCESS

Male and Female		
Age	**n**	**mmol/L**
Newborn	*	–10 to –2
Infant		–7 to –1
Child		–4 to +2
Thereafter		–3 to +3

Specimen Type(s)	Whole blood
Reference(s)	Behrman RE, ed. Nelson textbook of pediatrics, 14th ed. Philadelphia, PA: WB Saunders Company, 1992:1803.
Method(s)	Not described.
Comment(s)	*Numbers not given.

BILE ACIDS (TOTAL) (TOTAL 3-α-HYDROXY BILE ACIDS)

Male and Female		
Age	**n**	**µmol/L**
1 y	42	11–69
3 y	35	1–41
6 y	50	1–30
9 y	36	1–28
12 y	37	1–37
Adult	7	1–7

Specimen Type(s)	Serum
Reference(s)	McGraw C, Ellinor YM, Heubi JE. Reference ranges for total serum bile acids in infants. Clin Chem 1996;42:S307. (Abstract)
Method(s)	Nonradioactive enzymatic spectrophotometric method. "Enzabile," adapted for use on the Hitachi 705 analyzer (Boehringer Mannheim Diagnostics, Indianapolis, IN).
Comment(s)	Results are 2.5–97.5th percentiles. Population studied was a nonfasting, healthy population.

BILIRUBIN (CONJUGATED)

Male and Female				
Test	Age	n	µmol/L	mg/dL
1	Neonates	*	<10	<0.6
	>Neonates		<2	<0.1
2	Preterm infants			
	(1–6 d)	30	<10	<0.6
3	Neonates	**	<7	<0.4

Specimen Type(s)	1,2	Serum
	3	Plasma/serum
Reference(s)	1	Reference values and SI information. The Hospital for Sick Children, Toronto, Canada, 1993:359.
	2	Lockitch G, Halstead AC, Albersheim S, et al. Age- and sex-specific pediatric reference intervals for biochemistry analytes as measured with the Ektachem-700 analyzer. Clin Chem 1988;34:1622–5.
	3	Ghoshal AK, Soldin SJ. Evaluation of the Dade Behring Dimension RxL: integrated chemistry system-pediatric reference ranges. Clin Chim Acta 2003;331:135–46.
Method(s)	1,2	Vitros 700 (Ortho-Clinical Diagnostics, Raritan, NJ).
	3	Diazotized sulfanilic acid is formed by combining sodium nitrite and sulfanilic acid at low pH. Dade Behring Dimension RxL Analyzer (Dade Behring Inc., Newark, DE).
Comment(s)	1	*Numbers not given.
	2	Results are 97.5th percentile.
	3	**Reference ranges were obtained by comparing results from previously published data and using regression equations.

BILIRUBIN (TOTAL)

Male and Female				
Test	Age	n	μmol/L	mg/dL
1	Birth–1 d	*	<100	<5.8
	1–2 d		<140	<8.2
	3–5 d		<200	<11.7
	1 mo–adult		<17	<1.0
2	Bottle fed infants	2416**	<212	<12.4
	Breast fed infants	2416**	<253	<14.8
3	0–1 d	***	<87	<5.1
	1–2 d		<123	<7.2
	3–5 d		<176	<10.3
	1 mo–adult		<14	<0.8

Specimen Type(s)	1,2	Serum
	3	Plasma/serum
Reference(s)	1	Reference values and SI information. The Hospital for Sick Children, Toronto, Canada, 1993:359.
	2	Maisels MJ, Gifford K. Normal serum bilirubin levels in the newborn and the effect of breast-feeding. Pediatrics 1986;78:837–43.
	3	Ghoshal AK, Soldin SJ. Evaluation of the Dade Behring Dimension RxL: integrated chemistry system-pediatric reference ranges. Clin Chim Acta 2003;331:135–46.
Method(s)	1	Vitros 700 (Ortho-Clinical Diagnostics, Raritan, NJ).
	2	Modified diazo method on ACA III (DuPont Co., Clinical Systems Division, Wilmington, DE).
	3	Diazotized sulfanilic acid is formed by combining sodium nitrite and sulfanilic acid at low pH. Bilirubin in the sample, including the delta form (1), is solubilized by dilution in a mixture of caffeine/benzoate/acetate/EDTA. Dade Behring Dimension RxL Analyzer (Dade Behring Inc., Newark, DE). 1. Doumas BT, Wu T-W, Jendrzejczak. Delta bilirubin: absorption spectra, molar absorptivity, and reactivity in the diazo reaction. Clin Chem 1987;33:769–74.
Comment(s)	1	Should decrease to adult values by day 10 (breast-fed infants may take longer). Values in premature infants may be higher than in term infants and may reach peak concentrations at later times. *Numbers not provided.
	2	**2416 consecutive infants admitted to the well-baby nursery studied. This number includes both formula- and breast-fed infants. Results are 97th percentile. Breast feeding was significantly associated with hyperbilirubinemia.
	3	***Reference ranges were obtained by comparing results from previously published data and using regression equations.

C1 ESTERASE INHIBITOR

		Males		Females	
Test	**Age**	**n**	**mg/L**	**n**	**mg/L**
1	0–30 d	27	75–170	19	80–150
	31–182 d	39	105–279	49	105–229
	6 mo–<4 y	106	145–268	103	159–280
	4–<7 y	75	140–252	48	120–269
	7–<10 y	54	135–270	31	136–190
	10–<13 y	60	118–209	61	142–211
	13–<16 y	68	130–229	56	127–201
	16–18 y	35	130–216	47	140–200

Specimen Type(s)	Plasma/serum
Reference(s)	Soldin SJ, Hicks JM, Bailey J, et al. Pediatric reference ranges for estradiol and C1 esterase inhibitor. Clin Chem 1998;44:A17. (Abstract)
Method(s)	DiaSorin kit (DiaSorin Corp., Stillwater, MN).
Comment(s)	The study used plasma/serum from hospitalized patients and employed Chauvenet's criteria for removing outliers and a computerized approach adapted from the Hoffmann technique to obtain the 2.5–97.5th percentiles.

CALCIUM

		Male			Female		
Test	**Age**	**n**	**mg/dL**	**mmol/L**	**n**	**mg/dL**	**mmol/L**
1	0–7 d	293	7.3–11.4	1.83–2.85	259	7.5–11.3	1.88–2.83
	8–30 d	434	8.6–11.7	2.15–2.93	264	8.4–11.9	2.10–2.98
	31–90 d	371	8.5–11.3	2.13–2.83	265	8.0–11.1	2.00–2.78
	91–180 d	186	8.3–11.4	2.08–2.85	222	7.7–11.5	1.93–2.88
	181–365 d	429	7.7–11.0	1.93–2.75	362	7.8–11.1	1.95–2.7
2	0–5 d (<2.5 kg)	50*	7.9–10.7	1.96–2.66	50*	7.9–10.7	1.96–2.66
	1–3 y	50*	8.7–9.8	2.17–2.44	50*	8.7–9.8	2.17–2.44
	4–6 y	38*	8.8–10.1	2.19–2.51	38*	8.8–10.1	2.19–2.51
	7–9 y	72*	8.8–10.1	2.19–2.51	72*	8.8–10.1	2.19–2.51
	10–11 y	62*	8.9–10.1	2.22–2.51	62*	8.9–10.1	2.22–2.51
	12–13 y	73*	8.8–10.6	2.19–2.64	73*	8.8–10.6	2.19–2.64
	14–15 y	91*	9.2–10.7	2.29–2.66	91*	9.2–10.7	2.29–2.66
	16–19 y	107*	8.9–10.7	2.22–2.66	107*	8.9–10.7	2.22–2.66
3	1–30 d	62	8.5–10.6	2.12–2.64	66	8.4–10.6	2.10–2.64
	31–365 d	83	8.7–10.5	2.17–2.62	66	8.9–10.5	2.22–2.62
	1–3 y	126	8.8–10.6	2.19–2.64	119	8.5–10.4	2.12–2.59
	4–6 y	112	8.8–10.6	2.19–2.64	106	8.5–10.6	2.12–2.64
	7–9 y	117	8.7–10.3	2.17–2.57	107	8.5–10.3	2.12–2.57
	10–12 y	135	8.7–10.2	2.17–2.54	115	8.6–10.2	2.15–2.64
	13–15 y	109	8.5–10.2	2.12–2.54	110	8.4–10.0	2.10–2.50
	16–18 y	95	8.4–10.3	2.10–2.57	122	8.6–9.8	2.15–2.45

			Male		Female	
Test	**Age**	**n**	**mg/dL**	**mmol/L**	**mg/dL**	**mmol/L**
4	0–7 d	**	7.6–11.3	1.90–2.82	7.8–11.2	1.95–2.79
	8–30 d		8.8–11.6	2.20–2.89	8.6–11.8	2.15–2.94
	31–90 d		8.7–11.2	2.17–2.79	8.2–11.0	2.05–2.74
	91–180 d		8.5–11.3	2.12–2.82	8.0–11.4	2.00–2.84
	181–365 d		8.0–10.9	2.00–2.72	8.1–11.0	2.02–2.74
	1–3 y		8.9–9.9	2.22–2.47	8.9–9.9	2.22–2.47
	4–6 y		9.0–10.1	2.25–2.52	9.0–10.1	2.22–2.47
	7–9 y		9.0–10.1	2.25–2.52	9.0–10.1	2.22–2.47
	10–11 y		9.0–10.1	2.25–2.52	9.0–10.1	2.22–2.47
	12–13 y		9.0–10.6	2.25–2.64	9.0–10.6	2.22–2.64
	14–15 y		9.3–10.7	2.32–2.67	9.3–10.7	2.32–2.67
	16–19 y		9.0–10.7	2.25–2.67	9.0–10.7	2.25–2.67

Specimen Type(s)	1,3,4	Plasma/serum
	2	Serum
Reference(s)	1	Soldin SJ, Morse AS. Pediatric reference ranges for calcium and triglycerides in children <1 year old using the Vitros 500 analyzer. Clin Chem 1998;44:A16. (Abstract)
	2	Lockitch G, Halstead AC, Albersheim S, et al. Age- and sex-specific pediatric reference intervals for biochemistry analytes as measured with the Ektachem-700 analyzer. Clin Chem 1988;34:1622–5.
	3	Soldin SJ, Hicks JM, Bailey J, et al. Pediatric reference ranges for calcium on the Hitachi 747 Analyzer. Clin Chem 1997;43:S198. (Abstract)
	4	Ghoshal AK, Soldin SJ. Evaluation of the Dade Behring Dimension RxL: integrated chemistry system-pediatric reference ranges. Clin Chim Acta 2003;331:135–46.
Method(s)	1,2	Arsenazo III dye method. Vitros 700 (Ortho-Clinical Diagnostics, Raritan, NJ).
	3	Cresolphthalein complexone. Hitachi 747 Boehringer Mannheim reagents (Boehringer Mannheim Diagnostics, Indianapolis, IN).
	4	The calcium method is a modification of the calcium o-cresolphthalein complexone (OCPC) reaction. Dade Behring Dimension RxL Analyzer (Dade Behring Inc., Newark, DE).
Comment(s)	1,3	Study used hospitalized patients and a computerized approach to removing outliers. Values are 2.5–97.5th percentiles.
	2	From normal healthy children. Values are 2.5–97.5th percentiles.
		*No significant differences were found for males and females. These ranges were therefore derived from combined data.
	4	**Reference ranges were obtained by comparing results from previously published data and using regression equations.

CALCIUM, IONIZED

			Male		Female	
Test	**Age**	**n**	**mg/dL**	**mmol/L**	**mg/dL**	**mmol/L**
1	0–1 mo	207	3.9–6.0	1.0–1.5	3.9–6.0	1.0–1.5
	1–6 mo	96	3.7–5.9	0.95–1.5	3.7–5.9	0.95–1.5
2	1–19 y	*	4.9–5.5	1.22–1.37		
	20 y–adult		4.75–5.3	1.18–1.32		
	1–17 y				4.9–5.5	1.22–1.37
	18 y–adult				4.75–5.3	1.18–1.32

Specimen Type(s)	1,2	Whole blood
Reference(s)	1	Snell J, Greeley C, Colaco A, et al. Pediatric reference ranges for arterial pH, whole blood electrolytes and glucose. Clin Chem 1993:39;1173. (Abstract)
	2	Burritt MF, Slockbower JM, Forsman RW, et al. Pediatric reference intervals for 19 biologic variables in healthy children. Mayo Clinic Proceedings 1990:65;329–36.
Method(s)	1	Ciba Corning 288 Blood Gas System (Ciba Corning Diagnostics, East Walpole, MA).
	2	Ion selective electrode Radiometer ICA 1 (Radiometer America, Inc., Cleveland, OH).
Comment(s)	1	Study used hospitalized patients and a computerized approach to removing outliers. Values are 2.5–97.5th percentiles.
	2	From normal healthy children. Values are 2.5–97.5th percentiles.
		*See reference for numbers.

CARBON DIOXIDE (CO_2)

Male and Female		
Test	**Age**	**mmol/L**
1	0–1 wk	17–26
	1 wk–1 mo	17–27
	1–6 mo	17–29
	6 mo–1 y	18–29
	1 y	20–31
2	Infants	13–29
	Adults	24–30
3	0–1 wk	13–21
	1 wk–1 mo	13–22
	1–6 mo	13–23
	6 mo–1 y	14–23
	>1 y	16–25

Specimen Type(s)	1	Plasma
	2	Cord blood
	3	Plasma/serum
Reference(s)	1	Greeley C, Snell J, Colaco A, et al. Pediatric reference ranges for electrolytes and creatinine. Clin Chem 1993;39:1172. (Abstract)
	2	Chemistry analytes in arterial and venous umbilical cord blood. Clin Chem 1993;39:1041–4.
	3	Ghoshal AK, Soldin SJ. Evaluation of the Dade Behring Dimension RxL: integrated chemistry system-pediatric reference ranges. Clin Chim Acta 2003;331:135–46.
Method(s)	1	Vitros 700 (Ortho-Clinical Diagnostics, Raritan, NJ).
	2	Hitachi 737, Boehringer Mannheim reagents (Boehringer Mannheim, Montreal, Canada).
	3	The total carbon dioxide method uses a Severinghaus electrode designed to measure the liberated CO_2 from an acidified sample. Dade Behring Dimension RxL Analyzer (Dade Behring Inc., Newark, DE).
Comment(s)	1	Study used hospitalized patients and a computerized approach to removing outliers. N ≥100 in all of the above. Values are 2.5–97.5th percentiles.
	2	Study used 397 infants (209 girls and 188 boys delivered between 37 and 41 weeks gestation). Results are 2.5–97.5th percentiles.
	3	Reference ranges were obtained by comparing results from previously published data and using regression equations.

CARBON DIOXIDE, PARTIAL PRESSURE (pCO_2)

Male and Female			
Age	n	mmHg	kPa
Newborn	*	27–40	3.6–5.3
Infant		27–41	3.6–5.5
Thereafter		32–48	4.3–6.4

Specimen Type(s)	Whole blood
Reference(s)	Behrman RE, ed. Nelson textbook of pediatrics, 14th ed. Philadelphia, PA: WB Saunders Company, 1992:1804.
Method(s)	Not given.
Comment(s)	*Numbers not provided.

CARNITINE (TOTAL)

		Male		Female	
Test	Age	n	µmol/L	n	µmol/L
1	1–7 d	*	17–46	*	17–46
	2 y		24–66		24–66
	<2 y		37–89		30–73
2	1–12 mo	12	15–39	12	15–39
	1–7 y	27	18–37	27	18–37
	7–15 y	9	31–43	9	31–43

Specimen Type(s)	1	Serum
	2	Plasma
Reference(s)	1	Mayo Medical Laboratories test catalog, 1994:78.
	2	Bonnefont JP, Specola NP, Vassault A, et al. The fasting test in pediatrics: application to the diagnosis of pathological hypo- and hyperketotic states. Eur J Pediatr 1990;150:80–5.
Method(s)	1	Radioisotope enzymatic.
	2	Radioisotopic assay. See reference.
Comment(s)	*Note:*	Free carnitine normally comprises 60–80% of the total carnitine.
	1	The total carnitine minus the free carnitine gives the esterified carnitine.
		*Numbers not provided.
	2	The above results are 10–90th percentiles and refer to 15-h fasting values. For 20- and 24-h fasting values, see reference.

CERULOPLASMIN

		Male		Female	
Test	Age	n	mg/L	n	mg/L
1	0–5 d	73*	55–286	73*	55–286
	1–3 y	51*	264–506	51*	264–506
	4–6 y	39*	264–462	39*	264–462
	7–9 y	39*	264–440	39*	264–440
	10–13 y	36	242–396	45	253–473
	14–19 y	46	154–374	66	220–495
2	1–30 d	35	77–253	36	33–275
	31–365 d	119	154–484	87	154–429
	1–3 y	127	253–561	114	286–539
	4–6 y	99	286–561	81	264–539
	7–9 y	75	253–517	84	231–484
	10–12 y	69	209–506	90	209–484
	13–15 y	73	198–495	72	209–462
	16–18 y	49	198–451	73	220–495

Specimen Type(s)	1	Serum
	2	Plasma/serum
Reference(s)		The above values have been adjusted from those published to convert results to current IFCC units using the new IFCC guidelines/standards.
	1	Lockitch G, Halstead AC, Quigley G, et al. Age- and sex-specific pediatric reference intervals: study design and methods illustrated by measurement of serum proteins with the Behring LN nephelometer. Clin Chem 1988;34:1618–21.
	2	Soldin SJ, Hicks JM, Bailey J, et al. Pediatric reference ranges for β2-microglobulin and ceruloplasmin. Clin Chem 1997;43:S199. (Abstract)
Method(s)	1	Nephelometry using Behring antisera and Behring LN nephelometer (Behring Diagnostics, Hoechst Canada, Inc., Montreal, Canada).
	2	Behring nephelometer with Behring reagents (Behring Diagnostics, Westwood, MA).
Comment(s)	1	Healthy normal children. Results represent the 0.025–0.975 fractiles. Males and females had similar ranges from 0–9 y and numbers quoted are combined for these age ranges.
		*No significant differences were found for males and females. These ranges were therefore derived from combined data.
	2	Study used hospitalized patients and a computerized approach adapted from the Hoffmann technique. Values are the 2.5–97.5th percentiles.

CHLORIDE

Male and Female			
Test	Age	n	mmol/L
1	0–1 wk	≥100	96–111
	1 wk–1 mo	≥100	96–110
	1–6 mo	≥100	96–110
	6 mo–1 y	≥100	96–108
	>1 y	≥100	96–109
2	1–17 y	*	102–112
	18–Adult		100–108
3	0–7 d	**	97–108
	7–31 d		97–108
	1–6 mo		97–108
	6 mo–1 y		97–106
	>1 y		97–107

Specimen Type(s)	1	Plasma
	2	Serum
	3	Plasma/serum
Reference(s)	1	Greeley C, Snell J, Colaco A, et al. Pediatric reference ranges for electrolytes and creatinine. Clin Chem 1993;39:1172. (Abstract)
	2	Burritt MF, Slockbower JM, Forsman BS, et al. Pediatric reference intervals for 19 biologic variables in healthy children. Mayo Clinic Proceedings 1990;65:329–36.
	3	Ghoshal AK, Soldin SJ. Evaluation of the Dade Behring Dimension RxL: integrated chemistry system-pediatric reference ranges. Clin Chim Acta 2003;331:135–46.
Method(s)	1	Vitros (Ortho-Clinical Diagnostics, Raritan, NJ).
	2	Coulometric-Beckman Astra 8 (Beckman Instruments, Inc., Palo Alto, CA).
	3	The sodium, potassium, and chloride (Na/K/Cl) methods use indirect sample sensing with the QuikLYTE® Integrated Multisensor Technology (IMT) to develop an electrical potential proportional to the activity of each specific ion in the sample. The total carbon dioxide (TCO_2) method uses a Severinghaus electrode designed to measure the liberated CO_2 from an acidified sample. Dade Behring Dimension RxL Analyzer (Dade Behring Inc., Newark, DE).
Comment(s)	1	Study used hospitalized patients and a computerized approach to removing outliers. Values are 2.5–97.5th percentiles. N ≥100 in all of the above.
	2	From normal children healthy children. Values are 2.5–97.5th percentiles. *See reference for numbers.
	3	**Reference ranges were obtained by comparing results from previously published data and using regression equations.

CHOLESTEROL

		Male			Female		
Test	**Age**	**n**	**mg/dL**	**mmol/L**	**n**	**mg/dL**	**mmol/L**
1	0–1 mo	37	45–177	1.16–4.58	27	63–198	1.63–5.12
	2–6 mo	354	60–197	1.55–5.09	243	66–218	1.71–5.64
	7–12 mo	401	89–208	2.30–5.39	252	74–218	1.91–5.64
2	1–3 y	49*	44–181	1.15–4.70	49*	44–181	1.15–4.70
	4–6 y	38*	108–187	2.80–4.80	38*	108–187	2.80–4.80
	7–9 y	72*	112–247	2.90–6.40	72*	112–247	2.90–6.40
	10–11 y	28	125–230	3.25–5.95	34	127–244	3.30–6.30
	12–13 y	32	127–230	3.30–5.95	40	125–213	3.25–5.55
	14–15 y	39	106–224	2.75–5.80	50	130–213	3.35–5.55
	16–19 y	41	110–220	2.85–5.70	68	106–217	2.75–5.60
3	1–30 d	62	54–151	1.40–3.90	74	62–155	1.60–4.01
	31–182 d	77	81–147	2.09–3.80	75	62–141	1.60–3.65
	183–365 d	53	76–179	1.97–4.63	45	76–216	1.97–5.59
	1–3 y	136	85–182	2.20–4.71	111	108–193	2.79–4.99
	4–6 y	112	110–217	2.84–5.61	113	106–193	2.74–4.99
	7–9 y	124	110–211	2.84–5.46	104	104–210	2.69–5.43
	10–12 y	111	105–223	2.72–5.77	109	105–218	2.72–5.64
	13–15 y	126	91–204	2.35–5.28	105	108–205	2.79–5.30
	16–18 y	112	82–192	2.12–4.97	110	92–234	2.38–6.05

			Male		Female	
Test	**Age**	**n**	**mg/dL**	**mmol/L**	**mg/dL**	**mmol/L**
4	0–1 mo	**	38–174	0.98–4.50	56–195	1.45–5.04
	2–6 mo		53–194	1.37–5.02	59–216	1.53–5.59
	7–12 mo		83–205	2.15–5.30	68–216	1.76–5.59
	1–3 y		37–178	0.96–4.60	37–178	0.96–4.60
	4–6 y		103–184	2.66–4.76	103–184	2.66–4.76
	7–9 y		107–245	2.77–6.34	107–245	2.77–6.34
	10–11 y		120–228	3.10–5.90	122–242	3.16–6.26
	12–13 y		122–228	3.16–5.90	120–211	3.10–5.46
	14–15 y		101–222	2.61–5.74	125–211	3.23–5.46
	16–18y		105–218	2.72–5.64	101–215	2.61–5.56

Specimen Type(s)	1–4	Plasma/serum
Reference(s)	1	Soldin SJ, Rakotoarisoa FTS. Pediatric reference ranges for amylase and cholesterol on the Kodak Ektachem 500 in the first year of life. Clin Chem 1996;42:S308. (Abstract)
	2	Lockitch G, Halstead AC, Albersheim S, et al. Age- and sex-specific pediatric reference intervals for biochemistry analytes as measured with the Ektachem-700 analyzer. Clin Chem 1988;34:1622–5.
	3	Hicks JM, Bailey J, Beatey J, et al. Pediatric reference ranges for cholesterol. Clin Chem 1996;42:S307. (Abstract)
	4	Ghoshal AK, Soldin SJ. Evaluation of the Dade Behring Dimension RxL: integrated chemistry system-pediatric reference ranges. Clin Chim Acta 2003;331:135–46.
Method(s)	1,2	Cholesterol oxidase method. Vitros 700 (Ortho-Clinical Diagnostics, Raritan, NJ).
	3	Boehringer Mannheim reagents on the Hitachi 747 analyzer (Boehringer Mannheim Diagnostics, Indianapolis, IN).
	4	Cholesterol esterase catalyzes the hydrolysis of cholesterol esters to produce free cholesterol, which along with preexisting free cholesterol, is oxidized in a reaction catalyzed by cholesterol oxidase. Dade Behring Dimension RxL Analyzer (Dade Behring Inc., Newark, DE).
Comment(s)	1,3	Values are 2.5–97.5th percentiles. Study used hospitalized patients and a computerized approach to removing outliers.
	2	The study population was healthy children. Non-parametric methods were used to determine the reference values. The central 95% were used. *Males and females were not studied separately; n refers to the total number of males *and* females studied in each age group.
	4	**Reference ranges were obtained by comparing results from previously published data and using regression equations.

COMPLEMENT FRACTION — C_{1r}

Male and Female		
Age	n	mg/L
Term newborn cord blood	125	27–65
<18 y	31	25–140

Specimen Type(s)	Plasma (EDTA or citrate)
Reference(s)	Sonntag J, Brandenburg U, Polzehl D, et al. Complement System in healthy term newborns: Reference values in umbilical cord blood. Ped Develop Path 1998;1:131–5
Method(s)	Single radial immunodiffusion (Binding Site, Birmingham, UK).
Comment(s)	Studies were performed on cord blood obtained from 125 healthy term newborns. Results are 5–95th percentiles.

COMPLEMENT FRACTION — C_2

Male and Female		
Age	n	mg/L
Term newborn cord blood	125	12–24
<18 y	31	18–40

Specimen Type(s)	Plasma (EDTA or citrate)
Reference(s)	Sonntag J, Brandenburg U, Polzehl D, et al. Complement System in healthy term newborns: Reference values in umbilical cord blood. Ped Develop Path 1998;1:131–5.
Method(s)	Single radial immunodiffusion (Binding Site, Birmingham, UK).
Comment(s)	Studies were performed on cord blood obtained from 125 healthy term newborns. Results are 5–95th percentiles.

COMPLEMENT FRACTION — C_{3a}

Male and Female		
Age	**n**	**µg/L**
Term newborn cord blood	125	4–255
<18 y	31	0–161

Specimen Type(s)	Plasma (EDTA or citrate)
Reference(s)	Sonntag J, Brandenburg U, Polzehl D, et al. Complement system in healthy term newborns: Reference values in umbilical cord blood. Ped Develop Path 1998;1:131–5.
Method(s)	Enzyme immunoassay (EIA: Fa. Progen Biotechnik GmbH, Heidelberg, Germany).
Comment(s)	Studies were performed on cord blood obtained from 125 healthy term newborns. Results are 5–95th percentiles.

COMPLEMENT FRACTION — C_{3c}

		Male			Female		
Test	**Age**	**n**	**mg/dL**	**g/L**	**n**	**mg/dL**	**g/L**
1	0–6 mo	70	56–150	0.56–1.5	71	56–135	0.56–1.35
	6 mo–3 y	121	86–179	0.86–1.79	117	72–164	0.72–1.64
	4–6 y	66	101–191	1.01–1.91	54	96–144	0.96–1.44
	7–9 y	70	116–210	1.16–2.10	38	104–186	1.04–1.86
	10–12 y	82	111–161	1.11–1.61	73	110–191	1.10–1.91
	13–15 y	68	98–201	0.98–2.01	74	111–177	1.11–1.77
	16–18 y	36	114–188	1.14–1.88	57	92–180	0.92–1.80

			Male		Female	
Test	**Age**	**n**	**mg/dL**	**g/L**	**mg/dL**	**g/L**
2	0–6 mo	*	38–100	0.38–1.00	39–91	0.39–0.91
	6 mo–3 y		58–119	0.58–1.19	49–109	0.49–1.09
	4–6 y		68–127	0.68–1.27	65–122	0.65–1.22
	7–9 y		78–140	0.78–1.10	70–124	0.70–1.24
	10–12 y		75–107	0.75–1.07	74–127	0.74–1.27
	13–15 y		66–134	0.66–1.34	75–118	0.75–1.18
	16–18 y		77–125	0.77–1.25	62–120	0.62–1.20

	Male and Female			
Test	**Age**	**n**	**mg/dL**	**g/L**
3	Healthy infants	32	45–147	0.45–1.47
4	0–5 d	73	39–156	0.39–1.56
	1–19 y	334	77–143	0.77–1.43

Specimen Type(s)	1,3	Serum
	2,4	Plasma/serum
Reference(s)	The above values have been adjusted from those published in references 1, 3, and 4 to convert results to current IFCC units using the new IFCC guidelines/standards.	
	1	Soldin SJ, Hicks JM, Bailey J, et al. Pediatric reference ranges for complement factors C3c and C4. Clin Chem 1998;44:A14. (Abstract)
	2	Ghoshal AK, Soldin SJ. Evaluation of the Dade Behring Dimension RxL: integrated chemistry system-pediatric reference ranges. Clin Chim Acta 2003;331:135–46.
	3	Zilow G, Zilow EP, Burger R, et al. Complement activation in newborn infants with early onset infection. Ped Res 1993;34:199–203.
	4	Lockitch G, Halstead AC, Quigley G, et al. Age- and sex-specific pediatric reference intervals: study design and methods illustrated by measurement of serum proteins with the Behring LN nephelometer. Clin Chem 1988;34:1618–21.
Method(s)	1	Behring nephelometer using Behring kits (Behring Diagnostics Inc. Westwood, MA).
	2	The C_3 method is a quantitative turbidimetric assay using endpoint detection based on the precipitation of C_3 by its polyclonal antibody. C_3 from serum reacts with its polyclonal antibody to form immune complexes. Addition of polyethylene glycol accelerates the formation of these complexes. The resulting turbidity is proportional to the concentration of C_3 in the sample. Dade Behring Dimension RxL Analyzer (Dade Behring Inc., Newark, DE).
	3	Radial Immunodiffusion (Behring Diagnostics, Frankfurt, Germany).
	4	Nephelometric on Behring LN nephelometer (Behring Diagnostics, Hoechst Canada, Inc., Montreal).
Comment(s)	1	The study used serum/plasma from hospitalized patients and employed Chauvenet's criteria for removing outliers and a computerized approach adapted from the Hoffmann technique to obtain the 2.5–97.5th percentiles.
	2	*Reference ranges were obtained by comparing results from previously published data and using regression equations.
	3	C_3 reported in serum from normal neonates. Results are 0–100th percentiles.
	4	Healthy children and adults. Results are 0.025–0.975 fractiles.

COMPLEMENT FRACTION — C_4

Male and Female				
Test	Age	n	mg/dL	g/L
1	0–5 d	73	5–33	0.05–0.33
	1–19 y	334	7–40	0.07–0.40
	Adult	30	10–37	0.10–0.37

		Male			Female		
Test	Age	n	mg/dL	g/L	n	mg/dL	g/L
2	0–6 mo	68	9–28	0.09–0.28	71	10–25	0.10–0.25
	6 mo–3 y	121	14–48	0.14–0.48	116	14–42	0.14–0.42
	4–6 y	66	20–47	0.20–0.47	54	17–41	0.17–0.41
	7–9 y	69	17–35	0.17–0.35	38	16–41	0.16–0.41
	10–12 y	82	18–36	0.18–0.36	73	14–47	0.14–0.47
	13–15 y	68	17–51	0.17–0.51	74	16–37	0.16–0.37
	16–18 y	36	17–38	0.17–0.38	57	11–40	0.11–0.40

			Male		Female	
Test	Age	n	mg/dL	g/L	mg/dL	g/L
3	0–6 mo	*	13–30	0.13–0.30	14–28	0.14–0.28
	6 mo–3 y		17–48	0.17–0.48	17–43	0.17–0.43
	4–6 y		23–47	0.23–0.47	21–42	0.21–0.42
	7–9 y		21–37	0.21–0.37	20–42	0.20–0.42
	10–12 y		21–38	0.21–0.38	18–47	0.18–0.47
	13–15 y		21–51	0.21–0.51	20–38	0.20–0.38
	16–18 y		21–39	0.21–0.39	15–41	0.15–0.41

Specimen Type(s)	1	Serum
	2,3	Plasma/serum
Reference(s)		The above values have been adjusted from those published in references 1 and 2 to convert results to current IFCC units using the new IFCC guidelines/standards.
	1	Lockitch G, Halstead AC, Quigley G, et al. Age- and sex-specific pediatric reference intervals: study design and methods illustrated by measurement of serum proteins with the Behring LN nephelometer. Clin Chem 1988;34:1618–21.
	2	Soldin SJ, Hicks JM, Bailey J, et al. Pediatric reference ranges for complement factors C3c and C4. Clin Chem 1998;44:A14. (Abstract)
	3	Ghoshal AK, Soldin SJ. Evaluation of the Dade Behring Dimension RxL: integrated chemistry system-pediatric reference ranges. Clin Chim Acta 2003;331:135–46.
Method(s)	1,2	Nephelometric, Behring LN nephelometer (Behring Diagnostics, Hoechst Canada, Inc., Montreal, and Westwood, MA).
	3	The C_4 method is a quantitative turbidimetric assay using endpoint detection based on the precipitation of C_4 by its polyclonal antibody. Dade Behring Dimension RxL Analyzer (Dade Behring Inc., Newark, DE).
Comment(s)	1	Normal healthy children. Values provided are 0.025–0.975 fractiles.
	2	The study used serum/plasma from hospitalized patients and employed Chauvenet's criteria for removing outliers and a computerized approach adapted from the Hoffmann technique to obtain the 2.5–97.5th percentiles.
	3	*Reference ranges were obtained by comparing results from previously published data and using regression equations.

COMPLEMENT FRACTION — C_5

Male and Female		
Age	**n**	**mg/L**
Term newborn cord blood	125	64–127
Adult range	31	83–169

Specimen Type(s)	Plasma (EDTA or citrate)
Reference(s)	Sonntag J, Brandenburg U, Polzehl D, et al. Complement system in healthy term newborns: Reference values in umbilical cord blood. Ped and Develop Path 1998;1:131–5.
Method(s)	Single radial immunodiffusion (Binding Site, Birmingham, UK).
Comment(s)	Studies were performed on cord blood obtained from 125 healthy term newborns. Results are 5–95th percentiles.

COMPLEMENT FRACTION — C_{5a}

Male and Female		
Age	**n**	**µg/L**
Term newborn cord blood	125	0.11–1.19
Adult range	31	0.00–0.62

Specimen Type(s)	Plasma (EDTA or citrate)
Reference(s)	Sonntag J, Brandenburg U, Polzehl D, et al. Complement system in healthy term newborns: Reference values in umbilical cord blood. Ped and Develop Path 1998;1:131–5.
Method(s)	Specific Sandwich EIA (FA. Behring, Marburg, Germany).
Comment(s)	Studies were performed on cord blood obtained from 125 healthy term newborns. Results are 5–95th percentiles.

COMPLEMENT FRACTION — C_7

Male and Female		
Age	n	mg/L
Term newborn cord blood	125	32–89
Adult range	31	35–79

Specimen Type(s)	Plasma (EDTA or citrate)
Reference(s)	Sonntag J, Brandenburg U, Polzehl D, et al. Complement system in healthy term newborns: Reference values in umbilical cord blood. Ped Develop Path 1998;1:131–5.
Method(s)	Single radial immunodiffusion (Binding Site, Birmingham, UK).
Comment(s)	Studies were performed on cord blood obtained from 125 healthy term newborns. Results are 5–95th percentiles.

COMPLEMENT FRACTION — FACTOR D

Male and Female		
Age	n	mg/L
Term newborn cord blood	125	3.6–7.3
Adult range	31	2.7–5.4

Specimen Type(s)	Plasma (EDTA or citrate)
Reference(s)	Sonntag J, Brandenburg U, Polzehl D, et al. Complement system in healthy term newborns: Reference values in umbilical cord blood. Ped Develop Path 1998;1:131–5.
Method(s)	Single radial immunodiffusion (Binding Site, Birmingham, UK).
Comment(s)	Studies were performed on cord blood obtained from 125 healthy term newborns. Results are 5–95th percentiles.

COMPLEMENT FRACTION — FACTOR H

Male and Female		
Age	**n**	**mg/L**
Term newborn cord blood	125	178–296
Adult range	31	146–553

Specimen Type(s)	Plasma (EDTA or citrate)
Reference(s)	Sonntag J, Brandenburg U, Polzehl D, et al. Complement system in healthy term newborns: Reference values in umbilical cord blood. Ped Develop Path 1998;1:131–5.
Method(s)	Single radial immunodiffusion (Binding Site, Birmingham, UK).
Comment(s)	Studies were performed on cord blood obtained from 125 healthy term newborns. Results are 5–95th percentiles.

COMPLEMENT FRACTION — FACTOR I

Male and Female		
Age	**n**	**mg/L**
Term newborn cord blood	125	15–32
Adult range	31	32–88

Specimen Type(s)	Plasma (EDTA or citrate)
Reference(s)	Sonntag J, Brandenburg U, Polzehl D, et al. Complement system in healthy term newborns: Reference values in umbilical cord blood. Ped Develop Path 1998;1:131–5.
Method(s)	Single radial immunodiffusion (Binding Site, Birmingham, UK).
Comment(s)	Studies were performed on cord blood obtained from 125 healthy term newborns. Results are 5–95th percentiles.

COMPLEMENT FRACTION — PROPERDIN

Male and Female		
Age	**n**	**mg/L**
Term newborn cord blood	125	5.6–14.2
Adult range	31	24–50

Specimen Type(s)	Plasma (EDTA or citrate)
Reference(s)	Sonntag J, Brandenburg U, Polzehl D, et al. Complement system in healthy term newborns: Reference values in umbilical cord blood. Ped Develop Path 1998;1:131–5.
Method(s)	Single radial immunodiffusion (Binding Site, Birmingham, UK).
Comment(s)	Studies were performed on cord blood obtained from 125 healthy term newborns. Results are 5–95th percentiles.

COPPER

		Male			Female		
Test	Age	n	µg/dL	µmol/L	n	µg/dL	µmol/L
1	0–5 d	27*	9–46	1.4–7.2	27*	9–46	1.4–7.2
	1–5 y	77*	80–150	12.6–23.6	77*	80–150	12.6–23.6
	6–9 y	44*	84–136	13.2–21.4	44*	84–136	13.2–21.4
	10–14 y	36	80–121	12.6–19.0	23	82–120	12.9–18.9
	15–19 y	55	64–171	10.1–18.4	31	72–160	11.3–25.2

		Male and Female		
Test	Age	n	µg/dL	µmol/L
2	0–<0.5 y	13	38–104	5.9–16.3
	0.5–<1.0 y	18	24–152	3.8–23.8
	1.0–<2.0 y	15	76–193	11.9–30.3
	2.0–<4.0 y	23	87–187	13.7–29.3
	4.0–<6.0 y	19	56–191	8.8–30.0
	6.0–<10.0 y	25	117–181	18.4–28.4
	10.0–<14.0 y	21	87–182	13.7–28.5
	14.0–<18.0 y	17	75–187	11.7–29.3

Specimen Type(s)	1	Serum
	2	Plasma/serum
Reference(s)	1	Lockitch G, Halstead A, Wadsworth L, et al. Age- and sex-specific pediatric reference intervals for zinc, copper, selenium, iron, vitamins A and E, and related proteins. Clin Chem 1988;34:1625–8.
	2	Rükgauer M, Klein J, Kruse-Jarres J.D. Reference values for the trace elements copper, manganese, selenium, and zinc in the serum/plasma of children, adolescents, and adults. J Trace Elements Med Biol 1997;11:92–8.
Method(s)	1	Electrothermal atomic absorption spectroscopy with deuterium background correction. Varian GTA-95 (Varian Canada, Inc., Mississauga, Canada).
	2	Atomic absorption spectrophotometry with Zeeman background compensation. Perkin Elmer ETAAS, Zeeman 3030 (Uberlingen, Germany).
Comment(s)	1	The study population was healthy children. Non-parametric methods were used to determine the 0.025 and 0.975 fractiles. *No significant differences were found for males and females. These ranges were therefore derived from combined data.
	2	Study population was drawn from patients visiting the outpatient department or surgical or orthopedic ward for preoperative workup. Results are mean ±2 SDs. (2.5–97.5th percentiles).

CORTISOL

		Male			Female		
Test	**Age**	**n**	**µg/L**	**nmol/L**	**n**	**µg/L**	**nmol/L**
1	5th day	*	6–198	17–546	*	6–198	17–546
	2–12 mo		24–229	66–632		24–229	66–632
	2–13 y		25–229	69–632		25–229	69–632
	14–15 y		25–229	69–632		24–286	66–789
	16–18 y		24–286	66–789		24–286	66–789
2	5th day	**	6–156	16–431	**	6–156	16–431
	2–12 mo		20–181	54–499		20–181	54–499
	2–13 y		21–181	57–499		21–181	57–499
	14–15 y		21–181	57–499		20–274	54–756
	16–18 y		20–225	54–622		20–274	54–756
Male and Female							
		5–11 am			**5–11 pm**		
Test	**Age**	**n**	**µg/L**	**nmol/L**	**n**	**µg/L**	**nmol/L**
3	0–24 mo	33	10–340	28–938	25	10–300	28–828
	2–<11 y	31	10–330	28–911	23	10–240	28–662
	11–18 y	31	10–280	28–773	19	10–220	28–607
4	0–24 mo	***	11–330	30–910	***	11–290	30–800
	2–<11y		11–319	30–880		11–232	30–640
	11–18 y		11–272	30–750		11–203	30–560

Specimen Type(s)	1–3	Serum
	4	Plasma/serum
Reference(s)	1	Jonetz-Mentzel L, Wiedemann G. Establishment of reference ranges for cortisol in neonates, infants, children and adolescents. Eur J Clin Chem Biochem 1993;31:525–9.
	2	Murthy JN, Hicks JM, Soldin SJ. Evaluation of the Technicon Immuno I Random Access Immunoassay Analyzer and calculation of pediatric reference ranges for endocrine tests, T-uptake, and ferritin. Clin Biochem 1995;28:181–5.
	3	Soldin SJ, Murthy JN, Agarwalla PK, et. al. Pediatric reference ranges for creatine kinase, CKMB, troponin I, iron and cortisol. Clin Biochem 1999; 32:77–80.
	4	DPC IMMULITE® 1000 reference ranges used at Children's National Medical Center, Washington, DC.
Method(s)	1	TDx fluorescence polarization immunoassay (Abbott Laboratories, Abbott Park, IL).
	2	Bayer (Technicon) Immuno I analyzer with Bayer kits (Bayer Corp., Tarrytown, NY).
	3	Bayer Immuno I using Bayer reagents (Bayer Corp., Tarrytown, NY).
	4	DPC IMMULITE® 1000 analyzer (Diagnostic Products Corporation, Los Angeles, CA).
Comment(s)	1	687 normal healthy neonates, infants, children and adolescents. Values are 2.5 to 97.5th percentiles. Results were rounded off to the nearest whole integer.
		*See reference for numbers.
	2	Results represent 2.5–97.5th percentiles.
		**Numbers not provided.
	3	Study used hospitalized patients and a computerized approach adapted from the Hoffmann technique. Values are the 2.5–97.5th percentiles. Note that pm values are only moderately lower than am values.
	4	***Numbers not provided.
		Pediatric reference ranges were determined by using regression equations to transform the previously established ranges on the Bayer Immuno 1 (Bayer Corp., Tarrytown, NY) based on the Hoffmann technique to their corresponding values on the IMMULITE® 1000.

C-REACTIVE PROTEIN (CRP)

Male and Female			
Test	**Age**	**n**	**µg/L**
1	Cord serum	*	10–350
	Adult		68–8200
2	Cord serum	48	15–6000
3	Newborn	**	13–444
	Adult		86–10406

Specimen Type(s)	1,2	Serum
	3	Plasma/serum
Reference(s)	1	Burtis CA, Ashwood CR, eds. Tietz textbok of clinical chemistry. 2nd ed. WB Saunders Co., 1994; 2184.
	2	Shine B, Gould J, Campbell C, et al. Serum C-reactive protein in normal and infected neonates. Clin Chim Acta 1985;148:97–103.
	3	Ghoshal AK, Soldin SJ. Evaluation of the Dade Behring Dimension RxL: integrated chemistry system-pediatric reference ranges. Clin Chim Acta 2003;331:135–46.
Method(s)	1	Nephelometric.
	2	Double antibody magnetizable particle radioimmunoassay (1).
		1. Shine B, DeBeer FC, Pepys MB. Solid phase radioimmunoassay for human C-reactive protein. Clin Chim Acta 1981;117:13–23.
	3	The CRP method is based on a particle-enhanced turbidimetric immunoassay (PETIA) technique. Dade Behring Dimension RxL Analyzer (Dade Behring Inc., Newark, DE).
Comment(s)	1	*Numbers not provided.
	2	48 cord sera obtained from cord blood of normal neonates. Results are 5–95th percentiles. Values above 10,000 µg/L were found in neonates with infection.
	3	**Reference ranges used by manufacturer.

CREATINE KINASE

		Male		Female	
Test	**Age**	**n**	**U/L**	**n**	**U/L**
1	0–90 d	71	28–300	65	42–470
	3–12 mo	129	24–170	90	26–240
	13–24 mo	121	27–160	120	24–175
	2–10 y	245	30–150	231	24–175
	11–14 y	86	30–150	77	30–170
	15–18 y	56	33–145	111	27–140
2	1–30 d	113	2–183	77	2–134
	31–182 d	95	2–129	64	2–146
	183–365 d	58	2–143	49	18–138
	1–3 y	129	2–163	143	2–134
	4–6 y	133	18–158	112	8–147
	7–9 y	142	2–177	113	26–145
	10–12 y	148	6–217	101	6–137
	13–15 y	144	2–251	124	2–143
	16–18 y	114	2–238	146	13–144

			Male	Female
Test	**Age**	**n**	**U/L**	**U/L**
3	0–90 d	*	29–303	43–474
	3–12 mo		25–172	27–242
	13–24 mo		28–162	25–177
	2–10 y		31–152	25–177
	11–14 y		31–152	31–172
	15–18 y		34–147	28–142

Specimen Type(s)	1–3	Plasma/serum
Reference(s)	1	Soldin SJ, Murthy JN, Agarwalla PK, et al. Pediatric reference ranges for creatine kinase, CKMB, troponin I, iron and cortisol. Clin Biochem 1999;32:77–80.
	2	Soldin SJ, Hicks JM, Bailey J, et al. Pediatric reference ranges for creatine kinase and insulin-like growth factor 1. Clin Chem 1997;43:S199. (Abstract)
	3	Ghoshal AK, Soldin SJ. Evaluation of the Dade Behring Dimension RxL: integrated chemistry system-pediatric reference ranges. Clin Chim Acta 2003;331:135–46.
Method(s)	1	Vitros 500 (Ortho-Clinical Diagnostics, Raritan, NJ). Creatine phosphate to creatine.
	2	Hitachi 747 using Boehringer Mannheim reagents (Boehringer Mannheim Diagnostics, Indianapolis, IN).
	3	The ATP formed is used to phosphorylate glucose in a reaction catalyzed by hexokinase, and the resulting glucose-6-phosphate is oxidized by glucose-6-phosphate dehydrogenase with the simultaneous reduction of nicotinamide adenine dinucleotide phosphate. Dade Behring Dimension RxL Analyzer (Dade Behring Inc., Newark, DE).
Comment(s)	1,2	Study used hospitalized patients and a computerized adaptation of the Hoffmann technique. Values are 2.5–97.5th percentiles.
	3	*Reference ranges were obtained by comparing results from previously published data and using regression equations.

CREATINE KINASE ISOENZYMES (CPK ISOENZYMES)

Male and Female				
Test	Age	n	CKMB %	CKBB %
1	Cord blood	*		
	5–8 h		0.3–3.1	0.3–10.5
	24–33 h		1.7–7.9	3.6–13.4
	72–100 h		1.8–5.0	2.3–8.6
	Adult		1.4–5.4	5.1–13.3

Test	Age	n	CKMB 97.5th percentile µg/L
2	0–30 d	76	4.5
	31–90 d	45	4.8
	3–6 mo	88	1.9
	7–12 mo	47	1.7
	1–18 y	97	1.7
3	0–30 d	**	4.2
	31–90 d		4.5
	3–6 mo		1.8
	7–12 mo		1.7
	1–18 y		1.7

Specimen Type(s)	1	Serum
	2,3	Plasma/serum
Reference(s)	1	Behrman RE, ed. Nelson textbook of pediatrics, 14th ed. Philadelphia, PA: WB Saunders Company, 1992:1807.
	2	Soldin SJ, Murthy JN, Agarwalla PK, et al. Pediatric reference ranges for creatine kinase, CKMB, troponin I, iron and cortisol. Clin Biochem 1999;32:77–80.
	3	Ghoshal AK, Soldin SJ. Evaluation of the Dade Behring Dimension RxL: integrated chemistry system-pediatric reference ranges. Clin Chim Acta 2003;331:135–46.
Method(s)	1	Not provided.
	2	Bayer Immuno I with Bayer reagents (Bayer Corp., Tarrytown, NY).
	3	A one-step enzyme immunoassay based on the "sandwich" principle. Dade Behring Dimension RxL Analyzer (Dade Behring Inc., Newark, DE).
Comment(s)	1	*Numbers not given.
	2	Study used hospitalized patients and a computerized approach adapted from the Hoffmann technique. Values are the 97.5th percentiles.
	3	**Reference ranges were obtained by comparing results from previously published data and using regression equations.

CREATININE

Test	Age	n	mg/dL	μmol/L
		Male and Female		
1	0–1 wk	*	0.6–1.1	53–97
	1 wk–1 mo		0.3–0.7	27–62
	1–6 mo		0.2–0.4	18–35
	7–12 mo		0.2–0.4	18–35
	1–18 y		0.2–0.7	18–62
Test	**Age**	**n**	**mg/dL**	**mmol/L**
2	0–1 wk	**	0.7–1.2	0.06–0.11
	1 wk–1 mo		0.3–0.8	0.03–0.07
	1 mo–1 y		0.2–0.5	0.02–0.04
	1–9 y		0.2–0.8	0.02–0.07
	10–18 y		0.5–1.1	0.04–0.10

		Male			Female		
Test	**Age**	**n**	**mg/dL**	**μmol/L**	**n**	**mg/dL**	**μmol/L**
3	1–30 d	42	0.5–1.2	44–106	40	0.5–0.9	44–80
	31–365 d	62	0.4–0.7	35–62	59	0.4–0.6	35–53
	1–3 y	103	0.4–0.7	35–62	126	0.4–0.7	35–62
	4–6 y	129	0.5–0.8	44–71	116	0.5–0.8	44–71
	7–9 y	121	0.6–0.9	53–80	110	0.5–0.9	44–80
	10–12 y	125	0.6–1.0	53–88	117	0.6–1.0	53–88
	13–15 y	135	0.6–1.2	53–106	141	0.7–1.1	62–97
	16–18 y	106	0.8–1.4	71–123	114	0.8–1.2	71–106

Specimen Type(s)	1,3	Plasma
	2	Plasma/serum
Reference(s)	1	Greeley C, Snell J, Colaco A, et al. Pediatric reference ranges for electrolytes and creatinine. Clin Chem 1993:39:1172. (Abstract)
	2	Ghoshal AK, Soldin SJ. Evaluation of the Dade Behring Dimension RxL: integrated chemistry system-pediatric reference ranges. Clin Chim Acta 2003;331:135–46.
	3	Soldin SJ, Hicks JM, Bailey J, et al. Pediatric reference ranges for creatinine on the Hitachi 747 analyzer. Clin Chem 1997;43:S198. (Abstract)
Method(s)	1	Vitros (Ortho-Clinical Diagnostics, Raritan, NJ).
	2	The creatinine method employs a modification of the kinetic Jaffe reaction reported by Larsen. Dade Behring Dimension RxL Analyzer (Dade Behring Inc., Newark, DE).
	3	Jaffe method on Hitachi 747 Boehringer Mannheim reagents (Boehringer Mannheim Diagnostics, Indianapolis, IN).
Comment(s)	1	Study used hospitalized patients and a computerized approached to removing outliers.
		*n ≥100. Values are 2.5–97.5th percentiles.
	2	**Reference ranges were obtained by comparing results from previously published data and using regression equations.
	3	Study used hospitalized patients and a computerized approach to removing outliers. Values reported are 2.5–97.5th percentiles.

CREATININE (URINE)

Male and Female		
Age	**n**	**g/24h**
3–8 y	71	0.11–0.68
9–12 y	45	0.17–1.41
13–17 y	42	0.29–1.87
Adult	104	0.63–2.50

Specimen Type(s)	Urine
Reference(s)	Pediatric endocrine testing. Nichols Institute, 1993:35.
Method(s)	Kinetic, alkaline picrate.
Comment(s)	Results are 2.5–97.5th percentiles.

DEHYDROEPIANDROSTERONE (DHEA)

		Male			Female		
Test	**Age**	**n**	**ng/dL**	**nmol/L**	**n**	**ng/dL**	**nmol/L**
1	6–9 y	15	13–187	0.5–6.5	11	18–189	0.6–6.6
	10–11 y	17	31–205	1.1–7.1	6	112–224	3.9–7.8
	12–14 y	16	83–258	2.9–8.9	12	98–360	3.4–12.5
	Adults	*	180–1250	6.2–43.3	*	130–980	4.5–34.0
2	0–1 d	**	320–1100	11–39	**	460–1200	16–42
	1–<7 d		90–870	3–30		120–930	4–32
	7–28 d		45–580	1.5–20		90–580	3–20
	1–12 mo		9–290	0.3–10		17–170	0.6–6
	1–<4 y		12–90	0.4–3		20–45	0.7–1.6
	4–10 y						
	P_1		25–300	0.9–10		12–200	0.4–7.0
	P_2		50–580	1.8–20		60–1700	2–60
	P_3		130–640	4.5–22		125–1900	4.4–65
	P_4		190–730	6.5–25		170–1700	6–60
	P_5		230–730	8.0–25		220–810	7.5–28

		Male and Female		
Test	**Age**	**n**	**ng/dL**	**nmol/L**
3	Premature infants	***		
	26–28 w, day 4		236–3640	8–127
	31–35 w, day 4		80–3150	3–110
	Full-term infants			
	3 d		65–1250	2–44
	7–30 d[a]		50–760	2–27
	1–6 mo		26–385	1–13
	6–12 mo		20–100	1–3.5
	Prepubertal children			
	1–6 y		20–130	1–4.5
	6–8 y[b]		20–275	1–9.6
	8–10 y		31–345	1–12

	Male			Female[c]		
Test	Age	ng/dL	nmol/L	Age	ng/dL	nmol/L
3	Puberty			Puberty		
cont.	Tanner Stage[d]			Tanner Stage[d]		
	1 (<9.8 y)	31–345	1–12	1 (<9.2 y)	31–345	1–12
	2 (9.8–14.5 y)	110–495	4–17	2 (9.2–13.7 y)	150–570	5–20
	3 (10.7–15.4 y)	170–585	6–20	3 (10.0–14.4 y)	200–600	7–21
	4 (11.8–16.2 y)	160–640	6–22	4 (10.7–15.6 y)	200–780	7–27
	5 (12.8–17.3 y)	250–900	9–32	5 (11.8–18.6 y)	215–850	7–30
	n***			n***		

Specimen Type(s)	1,3	Serum
	2	Plasma/serum
Reference(s)	1	Pediatric endocrine testing. Nichols Institute, 1993:13.
	2	Soldin SJ, Rifai N, Hicks JM, eds. Biochemical basis of pediatric disease, 3rd ed. Washington, DC: AACC Press, 1998:233.
	3	Endocrinology expected values. Esoterix Endocrinology, Calabasas Hills, CA. ©2002 Esoterix, Inc. http://www.esoterix.com
		[a]Data adapted from De Peretti E, Forest MG. Unconjugated dehydroepiandrosterone plasma levels in normal subjects from birth to adolescence in humans: the use of a sensitive radioimmunoassay. J Clin Endocrinol Metab 1976;43:982.
		[d]Tanner JM, Whitehouse RH. Clinical longitudinal standards for height, weight, height velocity, and the stages of puberty. Arch Dis Childhood 1976;51:170.
Method(s)	1	Extraction, chromatography, radioimmunoassay.
	2	See reference.
	3	Not provided.
Comment(s)	1	Results are 2.5–97.5th percentiles.
		*Numbers not provided.
	2	**Numbers not provided. These reference ranges are for guidance only.
		P_1–P_5 refer to pubertal stages.
	3	***Numbers not provided.
		[b]Values begin to increase progressively at about six years of age prior to any physical evidence of puberty.
		[c]Day of cycle not determined for prepubertal females.
		[d]As used herein, Tanner stages in males encompass development of both pubic hair and genitalia. In females, each stage encompasses both pubic hair and breast development. While this expands the chronological age range for each stage of pubertal development, it results in a better correlation with hormonal values.

DEHYDROEPIANDROSTERONE SULFATE (DHEAS)

		Male			Female		
Test	**Age**	**n**	**µg/dL**	**µmol/L**	**n**	**µg/dL**	**µmol/L**
1	0–1 mo	56	9–316	0.2–8.6	41	15–261	0.4–7.1
	1–6 mo	69	3–58	0.1–1.6	55	<2–74	<0.1–2.0
	7–12 mo	40	<2–26	<0.1–0.7	28	<2–26	<0.1–0.7
	1–3 y	114	<2–15	<0.1–0.4	121	<2–22	<0.1–0.6
	4–6 y	95	<2–27	<0.1–0.7	86	<2–34	<0.1–0.9
	7–9 y	103	<2–60	<0.1–1.6	83	<2–74	<0.1–2.0
	10–12 y	95	5–137	0.1–3.7	55	3–111	0.1–3.0
	13–15 y	86	3–188	0.1–5.1	78	4–171	0.1–4.6
	16–18 y	62	26–189	0.7–5.1	69	10–237	0.3–6.4
2	1–5 mo	*	<148	<4	*	<147	<4
	6 mo–7 y		<18	<0.5		<37	<1.0
	8–9 y		<110	<3		<110	<3
	10–12 y		<221	<6		<295	<8
	13–19 y		110–442	3–12		37–442	1–12

	Male and Female			
Test	**Age**	**n**	**µg/dL**	**µmol/L**
3	Premature infants	**		
	26–28 w, day 4		123–882	3.3–23.8
	31–35 w, day 4		122–710	3.3–19.2
	Full-term infants			
	3 d		88–356	2.4–9.6
	1–12 mo		[a]	[a]
	Prepubertal children			
	1–6 y		<5–57	<0.1 –1.5
	6–8 y		9–72	0.2–1.9
	8–10 y		13–115	0.3–3.1

Male			Female[b]		
Age	**µg/dL**	**µmol/L**	**Age**	**µg/dL**	**µmol/L**
Puberty			Puberty		
Tanner Stage[c]			Tanner Stage[c]		
1 (<9.8 y)	13–83	0.3–2.3	1 (<9.2 y)	19–144	0.5–3.9
2 (9.8–14.5 y)	42–109	1.1–3.0	2 (9.2–13.7 y)	34–129	0.9–3.5
3 (10.7–15.4 y)	48–200	1.3–5.4	3 (10.0–14.4 y)	32–226	0.9–6.1
4 (11.8–16.2 y)	102–385	2.8–10.5	4 (10.7–15.6 y)	58–260	1.6–7.1
5 (12.8–17.3 y)	120–370	3.3–10.1	5 (11.8–18.6 y)	44–248	1.2–6.7
n**			**n****		

Specimen Type(s)	1	Plasma/serum
	2,3	Serum
Reference(s)	1	Soldin SJ, Godwin ID, Bailey J, et al. Pediatric reference ranges for DHEA sulfate. Clin Chem 1993;39:1171. (Abstract)
	2	Babalola AA, Ellis G. Serum dehydroepiandrosterone sulfate in a normal pediatric population. Clin Biochem 1985;18:184–9.
	3	Endocrinology expected values. Esoterix Endocrinology, Calabasas Hills, CA. ©2002 Esoterix, Inc. http://www.esoterix.com
		[c]Tanner JM, Whitehouse RH. Clinical longitudinal standards for height, weight, height velocity, and the stages of puberty. Arch Dis Childhood 1976;51:170.
Method(s)	1	DPC Coat-A-Count DHEA-SO_4 procedure (Diagnostic Products Corporation, Los Angeles CA).
	2	Used rabbit anti-DHA-3-hemisuccinyl-bovine-serum albumin antibody (Radioimmunoassay, Inc., Toronto, Canada). Clin Biochem 1985;18:184–9.
	3	Not provided.
Comment(s)	1	Study used hospitalized patients and a computerized approach to removing outliers. Values are 2.5–97.5th percentiles.
	2	Study used outpatients (131 boys and 143 girls) and provides the upper reference range.
		*See reference for appropriate numbers in each age interval.
	3	**Numbers not provided.
		[a]DHEA-S levels fall to a range of 5–111 μ/dL during the first month and decrease further to a range of 5–48 μ/dL by 6 months of age.
		[b]Day of cycle not determined for prepubertal females.
		[c]As used herein, Tanner stages in males encompass development of both pubic hair and genitalia. In females, each stage encompasses both pubic hair and breast development. While this expands the chronological age range for each stage of pubertal development, it results in a better correlation with hormonal values.

DEOXYCORTICOSTERONE (DOC)

		Male and Female	
Age	n	ng/dL	nmol/L
Premature infants	*		
26–28 wks, day 4		20–105	0.60–3.17
Newborn		[a]	[a]
Full-term infants			
1–12 mo		7–49	0.21–1.48
Prepubertal children			
2–10 y		2–34	0.06–1.03
Pubertal children, 8:00 a.m.		2–19	0.06–0.57

Specimen Type(s)	Serum
Reference(s)	Endocrinology expected values. Esoterix Endocrinology, Calabasas Hills, CA. ©2002 Esoterix, Inc. http://www.esoterix.com
Method(s)	Not provided.
Comment(s)	*Numbers not provided. [a]Levels are markedly elevated at birth and decrease rapidly during the first week to the range found in older infants.

11-DEOXYCORTISOL (COMPOUND S)

Male and Female				
Test	**Age**	**n**	**µg/dL**	**nmol/L**
1	6–9 y	11	0.01–0.07	0.3–2.0
	10–11 y	17	0.01–0.09	0.3–2.6
	12–14 y	12	0.01–0.05	0.3–1.4
	15–17 y	10	0.02–0.05	0.6–1.4
	Post metyrapone reference range	*	>5.0	>144
2	Premature infants	**		
	26–28 wks, day 4		0.110–1.376	3.2–39.6
	31–35 wks, day 4		0.048–0.579	1.4–16.7
	Full-term infants			
	3 days		0.013–0.147	0.4–4.2
	1–12 mo		<0.01–0.156	0.3–4.5
	Prepubertal children, 8:00 a.m.		0.020–0.155	0.6–4.5
	Pubertal children, 8:00 a.m.		0.012–0.158	0.3–4.6

Specimen Type(s)	1,2	Serum
Reference(s)	1	Pediatric endocrine testing. Nichols Institute, 1993:14.
	2	Endocrinology expected values. Esoterix Endocrinology, Calabasas Hills, CA. ©2002 Esoterix, Inc. http://www.esoterix.com
Method(s)	1	Extraction, chromatography, radioimmunoassay.
	2	Not provided.
Comment(s)	1	Results are 2.5–97.5th percentiles.
		*Numbers not provided.
	2	**Numbers not provided.

11-DEOXYCORTISOL (COMPOUND S, METYRAPONE TEST)

Male and Female			
Age	**n**	**μg/dL**	**nmol/L**
Children and adults[a]	*		
Baseline		<1	<28.9
Post metyrapone			
Single dose test		7–18	202.3–520.2
Multiple dose test		10–25	289.0–722.5

Specimen Type(s)	Serum
Reference(s)	Endocrinology expected values. Esoterix Endocrinology, Calabasas Hills, CA. ©2002 Esoterix, Inc. http://www.esoterix.com
Method(s)	Not provided.
Comment(s)	[a]Results in children are not significantly different than those found in adults. See Limal JM, Basmaciogullari, Rapaport R. Evaluation of single oral dose metyrapone tests in children with hypopituitarism. Acta Paed Scand 1976;65:177. *Numbers not provided.

DIHYDROTESTOSTERONE (DHT)

		Male		Female	
Age	**n**	**ng/dL**	**pmol/L**	**ng/dL**	**pmol/L**
Premature infants	*	10–53	345–1828	2–13	69–448
Full-term infants		5–60	172–2069	<2–15	<69–517
1–7 mo[a]		[b]	[b]	[c]	[c]
Prepubertal children		<3	<103	<3	<103

Male			Female		
Age	**ng/dL**	**pmol/L**	**Age**	**ng/dL**	**pmol/L**
Puberty			Puberty		
Tanner Stage[d]			Tanner Stage[d]		
1 (<9.8 y)	<3	<103	1 (<9.2 y)	<3	<103
2 (9.8–14.5 y)	3–17	103–586	2 (9.2–13.7 y)	5–12	172–414
3 (10.7–15.4 y)	8–33	276–1138	3 (10.0–14.4 y)	7–19	241–655
4 (11.8–16.2 y)	22–52	759–1793	4 (10.7–15.6 y)	4–13	138–448
5 (12.8–17.3 y)	24–65	828–2241	5 (11.8–18.6 y)	3–18	103–621
n*			n*		

Specimen Type(s)	Serum
Reference(s)	Endocrinology expected values. Esoterix Endocrinology, Calabasas Hills, CA. ©2002 Esoterix, Inc. http://www.esoterix.com [a]Data adapted from Pang S, Levine LS, Chow D, Sagiani F, Saenger P, New MI. Dihydrotestosterone and its relationship to testosterone in infancy and childhood. J Clin Endocrinol Metab 1979;48:821. [d]Tanner JM, Whitehouse RH. Clinical longitudinal standards for height, weight, height velocity, and the stages of puberty. Arch Dis Childhood 1976;51:170.
Method(s)	Not provided.
Comment(s)	[b]DHT decreases rapidly the first week, then increases to 12–85 ng/dL between 30–60 days. Levels then decrease gradually to prepubertal values by seven months. [c]Levels decrease during the first month to <3 ng/dL and remain there until puberty. [d]As used herein, Tanner stages in males encompass development of both pubic hair and genitalia. In females, each stage encompasses both pubic hair and breast development. While this expands the chronological age range for each stage of pubertal development, it results in a better correlation with hormonal values.

DOPAMINE (URINE)

Male and Female				
Test	Age	n	μg/g creatinine	mmol/mol creatinine
1	0–24 mo	24	<3000	<2.216
	2–4 y	37	<1533	<1.132
	5–9 y	40	<1048	<0.774
	10–19 y	41	< 545	<0.403
2	<1 y	18	240–1290	0.177–0.953
	1–4 y	24	80–1220	0.059–0.901
	4–10 y	23	220–720	0.162–0.532
	10–18 y	20	120–450	0.089–0.332

Specimen Type(s)	1,2	Urine
Reference(s)	1	Soldin SJ, Lam G, Pollard A, et al. High performance liquid chromatographic analysis of urinary catecholamines employing amperometric detection: reference values and use in laboratory diagnosis of neural crest tumors. Clin Biochem 1980;13:285–91.
	2	Rosano TG. Liquid chromatographic evaluation of age related changes in the urinary excretion of free catecholamines in pediatric patients. Clin Chem 1984;30:301–3.
Method(s)	1,2	HPLC with electrochemical detection.
Comment(s)	1	Study involved healthy children. Results quoted are 95th percentile.
	2	Urine was obtained from 85 pediatric patients (diagnosis of neoplasia excluded). Results are 0–100th percentiles.

EPINEPHRINE/ADRENALINE (PLASMA)

Male and Female				
Test	Age	n	pg/mL	nmol/L
1	2–10 d	21	36–401	0.2–2.2
	10 d–3 mo	10	55–201	0.3–1.1
	3–12 mo	14	55–438	0.3–2.4
	12–24 mo	13	36–639	0.2–3.5
	24–36 mo	8	18–438	0.1–2.4
	36 mo–15 y	20	18–456	0.1–2.5
2	30 min after birth	16	48–256	0.263–1.403
	2 h after birth	16	92–140	0.504–0.767
	3 h after birth	16	41–121	0.225–0.663
	12 h after birth	16	8–12	0.044–0.066
	24 h after birth	16	9–21	0.049–0.115
	48 h after birth	16	18–34	0.099–0.186

Specimen Type(s)	1,2	Plasma
Reference(s)	1	Candito M, Albertini M, Politano S, et al. Plasma catecholamine levels in children. J Chrom Biomed Appl 1993;617:304–7.
	2	Eliot RJ, Lam R, Leake RD, et al. Plasma catecholamine concentrations in infants at birth and during the first 48 hours of life. J Pediatrics 1980;96:311–5.
Method(s)	1	High performance liquid chromatography.
	2	Radioenzymatic method.
Comment(s)	1	Study population consisted of 86 healthy children (62 males, 24 females) aged 2 days to 15 years. Results are mean ±2 SDs.
	2	Study performed on 16 term vaginally delivered infants. Results are mean ±2 SDs.

EPINEPHRINE/ADRENALINE (URINE)

Male and Female				
Test	Age	n	µg/g creatinine	mmol/mol creatinine
1	0–24 mo	24	<75	<0.046
	2–4 y	37	<57	<0.035
	5–9 y	40	<35	<0.022
	10–19 y	41	<34	<0.021
2	0–12 mo	18	0–375	0–0.232
	1–4 y	24	0–82	0–0.051
	4–10 y	22	5–93	0.003–0.057
	10–18 y	20	3–58	0.001–0.027

Specimen Type(s)	1,2	Urine
Reference(s)	1	Soldin SJ, Lam G, Pollard A, et al. High performance liquid chromatographic analysis of urinary catecholamines employing amperometric detection: Reference values and use in laboratory diagnosis of neural crest tumors. Clin Biochem 1980;13:285–91.
	2	Rosano TG. Liquid chromatographic evaluation of age related changes in the urinary excretion of free catecholamines in pediatric patients. Clin Chem 1984;30:301–3.
Method(s)	1,2	HPLC with electrochemical detection.
Comment(s)	1	Study involved healthy children. Results quoted are 95th percentile.
	2	Urine was obtained from 85 pediatric patients. (Patients with a discharge diagnosis of neoplasia, endocrinopathy, or muscular dystrophy were not included in the study.) Results are 0–100th percentiles.

ERYTHROPOIETIN

	Male		Female	
Age	n	mIU/mL	n	mIU/mL
1–3 y	122	1.7–17.9	97	2.1–15.9
4–6 y	89	3.5–21.9	76	2.9–8.5
7–9 y	79	1.0–13.5	80	2.1–8.2
10–12 y	98	1.0–14.0	90	1.1–9.1
13–15 y	10	2.2–14.4	148	3.8–20.5
16–18 y	66	1.5–15.2	77	2.0–14.2

Specimen Type(s)	Plasma
Reference(s)	Krafte–Jacobs B, Williams J, Soldin SJ. Plasma erythropoietin reference ranges in children. J Pediatrics 1995;126:601–3.
Method(s)	ELISA (Quantikine™ IVD™, Human EPO Immunoassay, R and D Systems, Minneapolis, MN).
Comment(s)	Study used hospitalized patients and a computerized approach to removing outliers. Values are 2.5–97.5th percentiles.

ESTRADIOL

		Male			Female		
Test	**Age**	**n**	**pg/mL**	**pmol/L**	**n**	**pg/mL**	**pmol/L**
1	0–30 d	38	8.0–53.0	29–194	22	12.0–66.0	44–241.6
	31–182 d	51	1.0–12.0	3.7–44	31	1.0–11.0	3.7–40.3
	6 mo–3 y	116	0.3–6.1	1.1–22.3	100	0.4–7.9	1.5–28.9
	4–6 y	43	0.9–5.5	3.3–20.1	27	1.8–8.3	6.6–30.4
	7–9 y	40	0.3–9.6	1.1–35.1	18	1.3–7.4	4.8–27.1
	10–12 y	43	0.4–10.0	1.5–36.6	27	0.3–20.6	1.1–75.4
	13–15 y	33	0.5–9.6	1.8–35.1	39	0.3–33.2	1.1–121.5
	16–18 y	14	3.6–20.4	13.2–74.7	24	4.4–49.2	16.1–180.1
2	1–6 y	*	<15	<55	*	<15	<55
	7–10 y		<15	<55		<70	<257
	11–12 y		<40	<147		10–300	37–1100
	13–15 y		<45	<165		10–300	37–1100
	16–17 y		10–50	37–184		10–300	37–1100
3	0–30 d	**	<70	<257	**	26–83	95–305
	31–182 d		<26	<95		<25	<92
	6 mo–6 y		<22	<81		<22	<81
	7–9 y		<23	<84		<22	<81
	10–12 y		<24	<88		<35	<128
	13–15 y		<23	<84		<48	<176
	16–18 y		<35	<128		<65	<239

	Male			Female		
Test	**Age**	**pg/mL**	**pmol/L**	**Age**	**pg/mL**	**pmol/L**
4	Newborn[a,b] (1–7 d)	[c]	[c]	Newborn[a,b] (1–7 d)	[c]	[c]
	1–6 mo[a]	[d]	[d]	1–6 mo[a]	[e]	[e]
	Prepubertal children (1–10 y)	<10	<37	Prepubertal children (1–10 y)	<10	<37
	Puberty[b]			Puberty[b,f]		
	Tanner Stage[g]			Tanner Stage[g]		
	1 (<9.8 y)	5–11	18–40	1 (<9.2 y)	5–20	18–74
	2 (9.8–14.5 y)	5–16	18–59	2 (9.2–13.7 y)	10–24	37–88
	3 (10.7–15.4 y)	5–25	18–92	3 (10.0–14.4 y)	7–60	26–220
	4 (11.8–16.2 y)	10–36	37–132	4 (10.7–15.6 y)	21–85	77–312
	5 (12.8–17.3 y)	10–36	37–132	5 (11.8–18.6 y)	34–170	125–625
	n***			n***		

Specimen Type(s)	1,3	Plasma/serum
	2	Plasma
	4	Serum
Reference(s)	1	Soldin SJ, Hicks JM, Bailey J, et al. Pediatric reference ranges for estradiol and C1 esterase inhibitor. Clin Chem 1998; 44:A17. (Abstract)
	2	Adaptation of method of Abraham GE, Odell WD, Swerdloff RS, et al. Simultaneous radioimmunoassay of plasma FSH, LH, progesterone, 17-hydroxy-progesterone and estradiol-17 beta during the menstrual cycle. J Clin Endocrinol Metab 1972;34:312–8.
		Meites S, ed. Pediatric clinical chemistry, 3rd ed. Washington, DC: AACC Press, 1989:122–3.
	3	DPC IMMULITE® 1000 reference ranges used at Children's National Medical Center, Washington, DC.
	4	Endocrinology expected values. Esoterix Endocrinology, Calabasas Hills, CA. ©2002 Esoterix, Inc. http://www.esoterix.com
		[a]Bidlingmaier F, Versmold H, Knorr D. Sex differences in plasma estrogen concentrations in infancy. 21 Symp Dtsch Ges Endokrin Abstract 103. Acta Endoc Suppl 193, 1975;103.
		[b]Data adapted from Bidlingmaier F, Wagner-Barnack M, Butenandt O, Knorr D. Plasma estrogens in childhood and puberty under physiologic and pathologic conditions. Pediat Res 1973;7:901.
		[g]Tanner JM, Whitehouse RH. Clinical longitudinal standards for height, weight, height velocity, and the stages of puberty. Arch Dis Childhood 1976;51:170.
Method(s)	1	Immunoassay. ACS 180 using Chiron reagents (Chiron Diagnostics Corp., East Walpole, MA).
	2	Extraction followed by RIA.
	3	DPC IMMULITE® 1000 analyzer (Diagnostic Products Corporation, Los Angeles, CA).
	4	Not provided.

Comment(s)	1	The study used serum/plasma from hospitalized patients and employed Chauvenet's criteria for removing outliers and a computerized approach adapted from the Hoffmann technique to obtain the 2.5–97.5th percentiles.
	2	*See pp. 122–3 of *Pediatric Clinical Chemistry* for numbers.
	3	**Numbers not provided.
		Pediatric reference ranges were determined by using regression equations to transform the previously established ranges on the Bayer Immuno 1 (Bayer Corp., Tarrytown, NY) based on the Hoffmann technique to their corresponding values on the IMMULITE® 1000.
	4	***Numbers not provided.
		[c]Estradiol levels are markedly elevated at birth and fall rapidly during the first week to prepubertal values.
		[d]Levels increase to 1.0–3.2 ng/dL between 30 and 60 days, then decline to <1.5 ng/dL by six months.
		[e]Levels increase to 0.5–5.0 ng/dL between 30 and 60 days, then decline to <1.5 ng/dL during the first year.
		[f]Day of cycle not determined for pubertal females.
		[g]As used herein, Tanner stages in males encompass development of both pubic hair and genitalia. In females, each stage encompasses both pubic hair and breast development. While this expands the chronological age range for each stage of pubertal development, it results in a better correlation with hormonal values.

ESTRONE

Male			Female		
Age	ng/dL	pmol/L	Age	ng/dL	pmol/L
Newborn[a,b] (1–7 d)	[c]	[c]	Newborn[a,b] (1–7 d)	[c]	[c]
Prepubertal children (1–10 y)	<1.5	<56	Prepubertal children (1–10 y)	<1.5	<56
Puberty[b]			Puberty[b,d]		
Tanner Stage[e]			Tanner Stage[e]		
1 (<9.8 y)	0.5–1.7	19–63	1 (<9.2 y)	0.4–2.9	15–107
2 (9.8–14.5 y)	1.0–2.5	37–93	2 (9.2–13.7 y)	1.0–3.3	37–122
3 (10.7–15.4 y)	1.5–2.5	56–93	3 (10.0–14.4 y)	1.5–4.3	56–159
4 (11.8–16.2 y)	1.5–4.5	56–167	4 (10.7–15.6 y)	1.6–7.7	59–285
5 (12.8–17.3 y)	2.0–4.5	74–167	5 (11.8–18.6 y)	2.9–10.5	107–389
n*			**n***		

Specimen Type(s)	Serum
Reference(s)	Endocrinology expected values. Esoterix Endocrinology, Calabasas Hills, CA. ©2002 Esoterix, Inc. http://www.esoterix.com [a]Bidlingmaier F, Versmold H, Knorr D. Sex differences in plasma estrogen concentrations in infancy. 21 Symp Dtsch Ges Endokrin Abstract 103. Acta Endoc Suppl 193, 1975;103. [b]Data adapted from Bidlingmaier F, Wagner-Barnack M, Butenandt O, Knorr D. Plasma estrogens in childhood and puberty under physiologic and pathologic conditions. Pediat Res 1973;7:901. [e]Tanner JM, Whitehouse RH. Clinical longitudinal standards for height, weight, height velocity, and the stages of puberty. Arch Dis Childhood 1976;51:170.
Method(s)	Not provided.
Comment(s)	*Numbers not provided. [c]Values are strikingly elevated at birth then decrease rapidly during the first week to prepubertal levels. [d]Day of cycle not determined for pubertal females. [e]As used herein, Tanner stages in males encompass development of both pubic hair and genitalia. In females, each stage encompasses both pubic hair and breast development. While this expands the chronological age range for each stage of pubertal development, it results in a better correlation with hormonal values.

FERRITIN

		Male		Female	
Test	**Age**	**n**	**ng/mL (µg/L)**	**n**	**ng/mL (µg/L)**
1	1–5 y*	44	6–24	44	6–24
	6–9 y*	50	10–55	50	10–55
	10–14 y	31	23–70	40	6–40
	14–19 y	65	23–70	110	6–40
2	1–30 d	83	6–400	66	6–515
	1–6 mo	70	6–410	50	6–340
	7–12 mo	51	6–80	51	6–45
	1–5 y	82	6–60	90	6–60
	6–19 y	77	6–320	121	6–70
3	1–30 d	83	36–381	66	36–483
	1–6 mo	70	36–391	50	36–329
	7–12 mo	51	36–100	51	36–70
	1–5 y	82	36–84	90	36–84
	6–19 y	77	36–311	121	36–92
4	0–6 wk	**	Up to 400	**	Up to 400
	7 wk–365 d		10–95		10–95
	1–9 y		10–60		10–60
	10–18 y		10–300		10–70
5	1–30 d	***	47–438	***	47–554
	1–6 mo		47–449		47–379
	7–12 mo		47–120		47–85
	1–5 y		47–101		47–101
	6–19 y		47–358		47–110

Specimen Type(s)	1–3,5	Plasma/serum
	4	Serum
Reference(s)	1	Lockitch G, Halstead A, Wadsworth L, et al. Age- and sex-specific pediatric reference intervals for zinc, copper, selenium, iron, vitamins A and E, and related proteins. Clin Chem 1988;34:1625–8.
	2	Soldin SJ, Morales A, Albalos F, et al. Pediatric reference ranges on the Abbott IMx for FSH, LH, prolactin, TSH, T_4, T_3, free T_4, Free T_3, T-uptake, and ferritin. Clin Biochem 1995;28:603–6.
	3	Murthy JN, Hicks JM, Soldin SJ. Evaluation of the Technicon Immuno I Random Access Immunoassay Analyzer and calculation of pediatric reference ranges for endocrine tests, T-uptake, and ferritin. Clin Biochem 1995;28:181–5.
	4	Dugaw KA, Jack RM, Rutledge J. Pediatric reference ranges for ferritin and AFP on the Vitros ECi analyzer. Clin Chem 2001;47:A108–9. (Abstract)
	5	Ghoshal AK, Soldin SJ. Evaluation of the Dade Behring Dimension RxL: integrated chemistry system-pediatric reference ranges. Clin Chim Acta 2003;331:135–46.
Method(s)	1	Ferritin was measured by immunoassay and using ferrizyme reagent (Abbott Laboratories, Abbott Park, IL).
	2	IMx analyzer (Abbott Laboratories, Abbott Park, IL).
	3	Bayer Immuno I with Bayer reagents (Bayer Corp., Tarrytown NY).
	4	Chemiluminescent immunoassay, Vitros ECi (Ortho-Clinical Diagnostics, Raritan, NJ).
	5	The ferritin (FERR) method for the Dimension® RxL clinical chemistry system is a one-step enzyme immunoassay based on the "sandwich" principle. Dade Behring Dimension RxL Analyzer (Dade Behring Inc., Newark, DE).
Comment(s)	1	The study population was healthy children. Non-parametric methods were used to determine the 0.025 and 0.975 fractiles.
		*No significant differences were found for males and females. These ranges were therefore derived from combined data.
	2, 3	Study used hospitalized patients and a computerized approach adapted from the Hoffmann technique. Values are 2.5–97.5th percentiles.
	4	**Ages ranged from 1 h to 18 y with a total of 92 specimens.
	5	***Numbers not provided.
		Reference ranges were obtained by comparing results from previously published data and using regression equations.
	Note:	Ferritin levels of <6 µg/L are strongly suggestive of iron deficiency.

FOLIC ACID

		Male		Female	
Test	Age	n	nmol/L	n	nmol/L
1	0–1 y	111	16.3–50.8	73	14.3–51.5
	2–3 y	105	5.7–34.0	135	3.9–35.6
	4–6 y	154	1.1–29.4	104	6.1–31.9
	7–9 y	103	5.2–27.0	102	5.4–30.4
	10–12 y	105	3.4–24.5	90	2.3–23.1
	13–18 y	127	2.7–19.9	159	2.7–16.3
2	Newborn[a]	*	16–72	*	16–72
	After newborn period		4–20		4–20
	Adult[a,b,c]		10–63		10–63

Specimen Type(s)	1	Plasma
	2	Serum
Reference(s)	1	Hicks JM, Cook J, Godwin ID, et al. Vitamin B_{12} and folate: pediatric reference ranges. Arch Pathol Lab Med, 1993;117:704–6.
	2	[a]Behrman RE, Vaughan VC, eds. Nelson textbook of pediatrics. Philadelphia, PA: WB Saunders Company 1983:1827–60.
		[b]Nathan DG, Oski FA. Hematology of infancy and childhood, 4th ed. Appendix IX. Philadelphia, PA: WB Saunders Company, 1993.
		[c]Hall CA, Bardwell SA, Allen ES, et al. Variation in plasma folate levels among groups of healthy persons. Am J Clin Nutr 1975;28:854–7.
Method(s)	1	Radioimmunoassay Quantaphase (BioRad, Hercules, CA).
	2	See reference.
Comment(s)	1	Study used hospitalized patients and a computerized approach adapted from the Hoffmann technique to obtain the 2.5–97.5th percentiles.
	2.	*See references for particulars of assays used and numbers studied.

FOLLICLE STIMULATING HORMONE (FSH)

		Male		Female	
Test	Age	n	U/L (mIU/mL)	n	U/L (mIU/mL)
1	<2 y	110	0.2–1.8	33	0.2–6.6
	2–5 y	124	0.2–1.4	96	0.2–3.8
	6–10 y	99	0.2–1.3	155	0.2–2.7
	11–20 y	100	0.2–8.0	247	0.2–8.0
2	<2 y	110	0.4–2.1	33	0.4–7.1
	2–5 y	124	0.4–1.7	94	0.4–4.2
	6–10 y	99	0.4–1.6	136	0.4–3.0
	11–20 y	100	0.4–8.7	263	0.4–8.6
3	<2 y	*	0.1–0.8	*	0.1–5.9
	2–5 y		0.1–0.4		0.1–3.0
	6–10 y		0.1–0.3		0.1–1.7
	11–18 y		0.1–7.5		0.1–7.4

	Male			Female		
Test	Age	n	U/L (mIU/mL)	Age	n	U/L (mIU/mL)
4	1–9 y	**	0.0–5.0	1–2 y	**	0.0–8.0
	10–11 y		0.0–6.0	3–8 y		0.0–5.0
	13–18 y		0.0–10.0	9–11 y		0.0–10.0
				12–18 y		0.0–15.0

Specimen Type(s)	1–3	Plasma/serum
	4	Serum
Reference(s)	1	Soldin SJ, Morales A, Albalos F, et al. Pediatric reference ranges on the Abbott IMx for FSH, LH, prolactin, TSH, T_4, T_3, free T_4, free T_3, T-uptake, IgE and ferritin. Clin Biochem 1995;28:603–6.
	2	Murthy JN, Hicks JM, Soldin SJ. Evaluation of the Technicon Immuno I Random Access Immunoassay Analyzer and calculation of pediatric reference ranges for endocrine tests, T-uptake, and ferritin. Clin Biochem 1995;28:181–5.
	3	DPC IMMULITE® 1000 reference ranges used at Children's National Medical Center, Washington, DC.
	4	Dugaw KA, Jack RM, Rutledge J. Pediatric reference ranges for FSH and LH on the Vitros ECi analyzer. Clin Chem 2001;47:A108. (Abstract)
Method(s)	1	IMx (Abbott Laboratories, Abbott Park, IL).
	2	Technicon Immuno I with Bayer reagents (Bayer Corp., Tarrytown, NY).
	3	DPC IMMULITE® 1000 analyzer (Diagnostic Products Corporation, Los Angeles, CA).
	4	Chemiluminescent immunoassay, Vitros ECi (Ortho-Clinical Diagnostics, Raritan, NJ).
Comment(s)	1,2	Study used hospitalized patients and a computerized approach to removing outliers. Values are 2.5–97.5th percentiles.
	3	*Numbers not provided.
		Pediatric reference ranges were determined by using regression equations to transform the previously established ranges on the Bayer Immuno 1 (Bayer Corp., Tarrytown, NY) based on the Hoffmann technique to their corresponding values on the IMMULITE® 1000.
	4	**Ages ranged from 1 h to 18 y with a total of 99 specimens.

FREE FATTY ACIDS

Male and Female		
Age	**n**	**mmol/L**
1–1 2 mo	12	0.5–1.6
1–7 mo	27	0.6–1.5
7–15 y	9	0.2–1.1

Specimen Type(s)	Plasma
Reference(s)	Bonnefont JP, Specola NB, Vassault A, et al. The fasting test in pediatrics: application to the diagnosis of pathological hypo- and hyperketotic states. Eur J Pediatr 1990;150:80–5.
Method(s)	Standard enzymatic procedure. See reference.
Comment(s)	Results are 10–90th percentiles.

FRUCTOSAMINE

Male and Female			
Test	**Age**	**n**	**mmol/L**
1	0–3 y	33	1.56–2.27
	3–6 y	49	1.73–2.34
	6–9 y	35	1.82–2.56
	9–12 y	24	2.04–2.50
	12–15 y	29	2.02–2.63
2	Normal	203	0.174–0.286
	Controlled diabetics	80	0.210–0.421
	Uncontrolled diabetics	123	0.268–0.870

Specimen Type(s)	1,2	Serum
Reference(s)	1	De Schepper J, Derde MP, Goubert P, et al. Reference values for fructosamine concentrations in children's sera: influence of protein concentration, age and sex. Clin Chem 1988;34:2444–7.
	2	RoTAG™ Fructosamine (glycated protein) Assay from package insert Roche diagnostics kit.
Method(s)	1	Roche diagnostics kit on Cobas-Bio (Roche Diagnostic Systems, Inc., Branchburg, NJ).
	2	Roche diagnostics kit (Roche Diagnostic Systems, Inc., Branchburg, NJ).
Comment(s)	1	Results are 2.5–97.5th percentiles.
	2	Results using the new Roche kit using polylysine differ by a factor of approximately 10 from the old kit using DMF equivalents. Ages not provided in study.

GAMMA-GLUTAMYLTRANSFERASE (GGT)

		Male		Female	
Test	Age	n	U/L	n	U/L
1	1–182 d	109	12–122	67	15–132
	183–365 d	36	1–39	36	1–39
	1–12 y	488	3–22	391	4–22
	13–18 y	170	2–42	208	4–24
2*	1–7 d	137	25–148	102	19–131
	8–30 d	186	23–153	129	17–124
	1–3 mo	172	17–130	189	17–124
	4–6 mo	241	8–83	191	15–109
	7–12 mo	199	10–35	190	10–54
3*	1–3 y	50**	5–16	50**	5–16
	4–6 y	40**	8–18	40**	8–18
	7–9 y	80**	11–21	80**	11–21
	10–11 y	27	14–25	34	14–23
	12–13 y	31	14–37	49	12–21
	14–15 y	26	10–28	52	12–22
	16–19 y	40	9–29	61	9–23
4	1–7 d	***	25–168	***	18–148
	8–30 d		23–174		16–140
	1–3 mo		16–147		16–140
	4–6 mo		5–93		13–123
	7–12 mo		8–38		8–59
	1–3 y		2–15		2–15
	4–6 y		5–17		5–17
	7–9 y		9–20		9–20
	10–11y		12–25		12–23
	12–13 y		12–39		10–20
	14–15 y		8–29		10–22
	16–19 y		6–30		6–23

Specimen Type(s)	1,2,4	Plasma/serum
	3	Plasma
Reference(s)	1	Soldin SJ, Hicks JM, Bailey J, et al. Pediatric reference ranges for gamma-glutamyltransferase. Clin Chem 1997;43:S198. (Abstract)
	2	Soldin SJ, Savwoir TV, Guo Y. Pediatric reference ranges for gamma-glutamyltransferase and urea nitrogen during the first year of life on the Vitros 500 analyzer. Clin Chem 1997;43:S199. (Abstract)
	3	Lockitch G, Halstead AC, Albersheim S, et al. Age- and sex-specific pediatric reference intervals for biochemistry analytes as measured with the Ektachem-700 analyzer. Clin Chem 1988;34:1622–5.
	4	Ghoshal AK, Soldin SJ. Evaluation of the Dade Behring Dimension RxL: integrated chemistry system-pediatric reference ranges. Clin Chim Acta 2003;331:135–46.
Method(s)	1	Substrate used is L-α-glutamyl-3-carboxy-4-nitroanilide and rate of production of 5-amino-2-nitrobenzoate is measured on the Hitachi 747. Boehringer Mannheim reagents (Boehringer Mannheim Diagnostics, Indianapolis, IN).
	2,3	Vitros 500 (2) and 700 (3) (Ortho-Clinical Diagnostics, Raritan, NJ). α-Glutamyl-p-nitroanilide used as substrate.
	4	Gamma-glutamyl transferase catalyzes the transfer of the glutamyl moiety from gamma-glutamyl-3-carboxy-4-nitranilide to glycylglycine thereby releasing 5-amino-2-nitrobenzoate, which absorbs at 405 nm. Dade Behring Dimension RxL Analyzer (Dade Behring Inc., Newark, DE).
Comment(s)	1,2	Study used hospitalized patients and a computerized approach adapted from the Hoffmann technique. Values are 2.5–97.5th percentiles.
	3	The study population was healthy children. Non-parametric methods were used to determine the 0.025 and 0.975 fractiles.
		*The values originally published were multiplied by 0.8383 to accommodate changes made by the manufacturer that decreased values by ~16%.
		**Males and females were not studied separately.
	4	***Reference ranges were obtained by comparing results from previously published data and using regression equations.

GASTRIN

Male and Female		
Age	n	ng/L pg/mL
Newborns (1–12 d)	*	69–190
Infants (1.5–22 mo)		55–186
Prepubertal and pubertal children, fasting		
3–4 h		2–168
5–6 h		3–117
≥8 h		1–125

Specimen Type(s)	Serum
Reference(s)	Pediatric endocrine testing. Nichols Institute, 1993:20.
Method(s)	Radioimmunoassay.
Comment(s)	Results are 2.5–97.5th percentiles. *See reference for numbers.

GLOBULINS
TOTAL, CALCULATED

		Male			Female		
Test	**Age**	**n**	**g/dL**	**g/L**	**n**	**g/dL**	**g/L**
1	1–182 d	58	1.3–2.4	13–24	51	1.3–2.1	13–24
	183–365 d	29	1.7–3.0	17–30	42	1.2–2.4	12–24
	1–3 y	134	2.1–3.4	21–34	186	1.8–3.3	18–33
	4–6 y	139	2.2–3.4	22–34	114	2.0–3.6	20–36
	7–9 y	118	2.2–3.6	22–36	107	2.2–3.4	22–34
	10–12 y	9	2.4–3.4	24–34	82	2.3–3.3	23–33
	13–15 y	7	2.1–3.7	21–37	83	2.4–3.8	24–38
	16–18 y	5	2.0–3.9	20–39	54	2.6–3.6	26–36
2	<1 y	104*	0.4–3.7	4–37	104*	0.4–3.7	4–37
	1–3 y	247*	1.6–3.5	16–35	247*	1.6–3.5	16–35
	4–9 y	694*	1.9–3.4	19–34	694*	1.9–3.4	19–34
	10–49 y	17905*	1.9–3.5	19–35	17905*	1.8–3.5	18–35

Specimen Type(s)	1	Plasma
	2	Serum
Reference(s)	1	Hicks JM, Bjorn S, Beatey J, et al. Pediatric reference ranges for albumin, globulin and total protein on the Hitachi 747. Clin Chem 1995;41:S93. (Abstract)
	2	Appleton C. Queensland Medical Laboratory, Brisbane, Australia.
		Meites, S, ed. Pediatric clinical chemistry, 3rd ed. Washington, DC: AACC Press, 1989:132.
Method(s)	1	Calculated (Total protein – albumin).
	2	Calculated. Obtained by subtracting albumin values from total protein values obtained on the Technicon SMAC II Analyzer (Technicon Instruments Corp., Tarrytown, NY).
Comment(s)	1	Study used hospitalized patients and a computerized approach adapted from the Hoffmann technique. Values are 2.5–97.5th percentiles.
	2	Values are 2.5–97.5th percentiles.
		*Numbers listed include males and females.

GLUCAGON

Male and Female		
Age	**n**	**ng/L pg/mL**
Cord blood	6	0–215
3 d	12	0–1750
4–14 y	9	0–148

Specimen Type(s)	Plasma
Reference(s)	Pediatric endocrine testing. Nichols Institute, 1993:20.
Method(s)	Extraction, radioimmunoassay.
Comment(s)	Results are 2.5–97.5th percentiles.

GLUCOSE

Male and Female				
Test	**Age**	**n**	**mg/dL**	**mmol/L**
1	0–1 mo	207	55–115	3.1–6.4
	1–6 mo	96	57–117	3.2–6.5
2	Outside the neonatal period	482*	70–126	3.9–7.0

			Male		Female	
Test	**Age**	**n**	**mg/dL**	**mmol/L**	**mg/dL**	**mmol/L**
3	0–1 d	**	36–110	2.00–6.11	36–89	2.00–4.94
	1–7 d		47–110	2.61–6.11	47–110	2.61–6.11
	>7 d		54–117	3.00–6.49	54–117	3.00–6.49

Specimen Type(s)	1	Whole blood
	2	Serum
	3	Plasma/serum
Reference(s)	1	Snell J, Greeley C, Colaco A, et al. Pediatric reference ranges for arterial pH whole blood electrolytes and glucose. Clin Chem 1993;39:1173. (Abstract)
	2	Lockitch G, Halstead AC, Albersheim S, et al. Age- and sex-specific pediatric reference intervals for biochemistry analyses as measured with the Ektachem-700 analyzer. Clin Chem 1988;34:1622–5.
	3	Ghoshal AK, Soldin SJ. Evaluation of the Dade Behring Dimension RxL: integrated chemistry system-pediatric reference ranges. Clin Chim Acta 2003;331:135–46.
Method(s)	1	Glucose oxidase YSI 2300 (Yellow Springs Instruments, Yellow Springs, OH).
	2	Glucose oxidase, Vitros 700 (Ortho-Clinical Diagnostics, Raritan, NJ).
	3	The glucose method is an adaptation of the hexokinase-glucose-6-phosphate dehydrogenase method. Dade Behring Dimension RxL Analyzer (Dade Behring Inc., Newark, DE).
Comment(s)	1	Results are 2.5–97.5th percentiles. The 0–1 mo group includes an appreciable number of premature infants.
	2	Results are 2.5–97.5th percentiles. *No significant differences were found for males and females. These ranges were therefore derived from combined data.
	3	**Numbers not provided. Reference ranges were obtained by comparing results from previously published data and using regression equations.

GLUCOSE (URINE)

Male and Female			
Age	n	mg/dL	mmol/L
All ages	*	6–19	0.33–1.05

Specimen Type(s)	Urine
Reference(s)	Ghoshal AK, Soldin SJ. Evaluation of the Dade Behring Dimension RxL: integrated chemistry system-pediatric reference ranges. Clin Chim Acta 2003;331:135–46.
Method(s)	The glucose method is an adaptation of the hexokinase-glucose-6-phosphate dehydrogenase method. Dade Behring Dimension RxL Analyzer (Dade Behring Inc., Newark, DE).
Comment(s)	*Numbers not provided. Reference ranges were obtained by comparing results from previously published data and using regression equations.

GLUCOSE CEREBROSPINAL FLUID (CSF GLUCOSE)

Test	Age	n	Male and Female	
1	All ages	*	60–80% of blood glucose	
Test	Age	n	mg/dL	mmol/L
2	All ages	**	41–84	2.28–4.66

Specimen Type(s)	1,2	Cerebrospinal fluid
Reference(s)	1	Meites S, ed. Pediatric clinical chemistry, 2nd ed. Washington, DC: AACC Press, 1981:216–8.
	2	Ghoshal AK, Soldin SJ. Evaluation of the Dade Behring Dimension RxL: integrated chemistry system-pediatric reference ranges. Clin Chim Acta 2003;331:135–46.
Methods	2	The glucose method is an adaptation of the hexokinase-glucose-6-phosphate dehydrogenase method. Dade Behring Dimension RxL Analyzer (Dade Behring Inc., Newark, DE).
Comment(s)	1	Glucose is measured in CSF primarily in the diagnosis of bacterial meningitis. CSF glucose should be compared to blood glucose for interpretation. *Number not provided.
	2	**Numbers not provided. Reference ranges were obtained by comparing results from previously published data and using regression equations.

GLUTATHIONE PEROXIDASE ACTIVITY

Male and Female		
Age	**n**	**U/L**
Newborn (term)	*	180–890
1–5 y		554–985
6–15 y		567–1153
16 y–adult		780–1269

Specimen Type(s)	Serum
Reference(s)	Lockitch G. Trace elements in pediatrics. JIFCC 1996:9;46–51.
Method(s)	Enzymatic.
Comment(s)	*Numbers of individuals included in each age range not provided. In selenium deficiency activity of glutathione peroxidase in plasma/serum reflects selenium status.

GROWTH HORMONE

Male and Female			
Test	**Age**	**n**	**µg/L**
1	1 mo–puberty	*	3–5
	After stimulation		>5
2	After stimulation by exercise, arginine, or insulin	**	>5
3	0–6 d	***	6–56
	7–30 d		6–38
	1–12 mo		<6
	1–17 y		<6
	After stimulation		>6

Specimen Type(s)	1,2	Serum
	3	Plasma/serum
Reference(s)	1	Soldin SJ, Rifai N, Hicks JM, eds. Biochemical basis of pediatric disease, 3rd ed. Washington, DC: AACC Press, 1998:157–79.
	2	Reference values and SI unit information. The Hospital for Sick Children, Toronto, Canada, 1993:369.
	3	DPC IMMULITE® 1000 reference ranges used at Children's National Medical Center, Washington, DC.
Method(s)	1	Not given.
	2	RIA.
	3	DPC IMMULITE® 1000 analyzer (Diagnostic Products Corporation, Los Angeles, CA).
Comment(s)	1	*Numbers not provided.
	2	**Numbers not provided.
		Differential diagnosis of short stature, slow growth or evaluation of pituitary function. Random samples have little diagnostic value.
	3	***Numbers not provided.

GROWTH HORMONE (URINE)

		Male			Female		
Test	**Age**	**n**	**ng/24h**	**ng/g creatinine**	**n**	**ng/24h**	**ng/g creatinine**
1	2.2–13.3 y Tanner 1	*	0.4–6.3	0.9–12.3	*	0.4–6.3	0.9–12.3
	10.3–14.6 y Tanner 2		0.8–12.0	1.0–14.1		0.8–12.0	1.0–14.1
	11.5–15.3 y Tanner 3		1.7–20.4	1.9–17.0		1.7–20.4	1.9–17.0
	12.7–17.1 y Tanner 4		1.5–18.2	1.3–14.4		1.5–18.2	1.3–14.4
	13.5–19.9 y Tanner 5		1.2–14.5	0.8–11.0		1.2–14.5	0.8–11.0
	Adults		0.6–20.9	0.4–15.9		0.6–20.9	0.4–15.9
2	3–4 y	**		>6.5	**		>9.1
	5–6 y			>8.7			>8.9
	7–8 y			>6.8			>5.9
	9–10 y			>5.6			>8.9
	11–12 y			>6.2			>9.3
	13–14 y			>10.5			>6.3
	15–16 y			>3.8			>5.1
	17–18 y			>3.1			>3.1
	19–20 y			>1.9			>1.9

Specimen Type(s)	1, 2	Urine
Reference(s)	1	Main K, Philips M, Jergensen M, et al. Urinary growth hormone excretion in 657 healthy children and adults. Normal values, inter- and intraindividual variations. Horm Res 1991;36:174–82.
	2	Nukada O, Moriwake T, Kanzaki S, et al. Age-related changes in urinary growth hormone level and its clinical application. Acta Paediatr Jpn 1990;32:32–8.
Method(s)	1	ELISA, antipituitary hGH (Nanormon® and recombinant hGH Norditropin®) (Novo Nordisk A/S, Gentofte, Denmark).
	2	EIA (Sumitomo Chemical Co., Ltd., Takarazuka, Japan).
Comment(s)	1	547 healthy children and 110 healthy adults.
		*See reference for numbers. Results are 2.5–97.5th percentiles.
	2	Study evaluated urinary GH secretion in 270 normal subjects aged 3–20 years. Results provided are the *lower* limit of normal (Mean–2 SDs).
		**See reference for numbers.

HEMOGLOBIN A_1c

Male and Female			
Test	Age	n	%
1	Normal individuals	*	4–7
	Pregnant women		5–8
	Old age		6–9
	Stable diabetics		8–10
	Young or unstable diabetics		8–18
2	Normal individuals	**	3.4–6.1
3	Healthy non-diabetics	***	4–6
	Diabetics with average control		9–10
4	Normal	****	3.7–6.2

Specimen Type(s)	1–3	Whole blood
	4	Plasma/serum
Reference(s)	1	Kellen JA. Disorders of carbohydrate metabolism. In: AG Gornall, ed. Applied biochemistry of clinical disorders, 2nd ed. Philadelphia, PA: JB Lippincott Company, 1986:379–402.
	2	Children's National Medical Center, Washington, DC. Unpublished data.
	3	Reference values and SI unit information. The Hospital for Sick Children, Toronto, Canada, 1993.
	4	Ghoshal AK, Soldin SJ. Evaluation of the Dade Behring Dimension RxL: integrated chemistry system-pediatric reference ranges. Clin Chim Acta 2003;331:135–46.
Method(s)	1	Not provided.
	2	Immunochemical technique DCA 2000 Hemoglobin A_1C System. (Miles Diagnostics Division, Elkhart, IN).
	3	High-performance liquid chromatography.
	4	Uses Dade Behring reagents. Dade Behring Dimension RxL Analyzer (Dade Behring Inc., Newark, DE).
Comment(s)	1	*Numbers not provided.
	2	**Numbers not provided. Hemoglobin is glycated in proportion to the average blood glucose concentration over the life span of the red cell (90–120 d). Hemoglobin A_1c allows tracking of metabolic control in individual patients over time.
	3	***Numbers not provided.
	4	****Numbers not provided.
		Reference ranges were obtained by comparing results from previously published data and using regression equations.

HIGH-DENSITY LIPOPROTEIN CHOLESTEROL (HDL CHOLESTEROL)

		Male			Female		
Test	**Age**	**n**	**mg/dL**	**mmol/L**	**n**	**mg/dL**	**mmol/L**
1	1–9 y	50*	35–82	0.91–2.12	50*	35–82	0.91–2.12
	10–13 y	56*	36–84	0.93–2.17	56*	36–84	0.93–2.17
	14–19 y	60*	35–65	0.91–1.68	60*	35–65	0.91–1.68
2	5 y	**	37–72	0.96–1.86	**	34–69	0.88–1.78
	10 y		31–70	0.80–1.81		34–69	0.88–1.78
	15 y		29–61	0.75–1.58		34–69	0.88–1.78
	20 y		29–58	0.75–1.50		36–70	0.93–1.81
3	0–24 mo	43***	8–61	0.21–1.58	43***	8–61	0.21–1.58
	2–<7 y	95	23–70	0.60–1.81	71	12–64	0.31–1.66
	7–<12 y	99	25–79	0.65–2.05	133	23–80	0.60–2.07
	12–<16 y	199	19–76	0.49–1.97	181	25–83	0.65–2.15
	16–<19 y	91	26–75	0.67–1.94	157	21–77	0.54–1.99
4	0–24 mo	****	12–60	0.31–1.55	****	12–60	0.31–1.55
	2–<7 y		26–68	0.67–1.76		16–62	0.41–1.61
	7–<12 y		28–76	0.73–1.97		26–77	0.67–1.99
	12–<16 y		22–73	0.57–1.89		28–79	0.73–2.05
	16–<19 y		28–72	0.73–1.86		24–74	0.62–1.92

Specimen Type(s)	1	Serum
	2	Plasma
	3,4	Plasma/serum
Reference(s)	1	Lockitch G, Halstead AC, Albersheim S, et al. Age- and sex-specific pediatric reference intervals for biochemistry analytes as measured with the Ektachem-700 analyzer. Clin Chem 1988;34:1622–5.
	2	Kottke BA, Moll PP, Michels VV, et al. Levels of lipids, lipoproteins and apolipoproteins in a defined population. Mayo Clin Proc 1991;66:1198–1208.
	3	Murthy JN, Soldin SJ. Pediatric reference ranges for HDL-cholesterol on the Vitros 500 analyzer. Clin Chem 1999; 45; A23. (Abstract)
	4	Ghoshal AK, Soldin SJ. Evaluation of the Dade Behring Dimension RxL: integrated chemistry system-pediatric reference ranges. Clin Chim Acta 2003;331:135–46.
Method(s)	1,3	Dextran sulfate precipitation/cholesterol oxidase. Vitros 700 (1) and 500 (3) (Ortho-Clinical Diagnostics, Raritan, NJ).
	2	LDL and VLDL precipitated using polyethylene glycol 6000. Supernatant used to determine HDL-cholesterol by an enzymatic method. See reference.
	4	The HDL cholesterol assay is a homogeneous method for directly measuring HDL levels without the need for off-line pretreatment or centrifugation steps. Dade Behring Dimension RxL Analyzer (Dade Behring Inc., Newark, DE).
Comment(s)	1	Healthy children. Results are 2.5–97.5th percentiles.
		*No significant differences were found for males and females. These ranges were therefore derived from combined data.
	2	Results are 5–95th percentiles read off graphs in reference.
		**See reference for numbers.
	3	***Male and female combined. The study used hospitalized patients and employed Chauvenet's criteria for removing outliers and a computerized approach adapted from the Hoffmann technique to obtain the 2.5–97.5th percentiles. The 2.5th and 97.5th percentiles are lower during the first 2 y of life. For children >2 y, the 2.5th percentiles are lower than those previously described by Lockitch, et al. Clin Chem 1988;34:1622–5.
	4	****Reference ranges were obtained by comparing results from previously published data and using regression equations.

HOMOCYSTEINE (TOTAL)

Male and Female		
Age	**n**	**µmol/L**
2 mo–10 y	105	3.3–8.3
11–15 y	59	4.7–10.3
16–18 y	31	4.7–11.3

Specimen Type(s)	Plasma
Reference(s)	Vilaseca MA, Moyano D, Ferrer I, et al. Total homocysteine in pediatric patients. Clin Chem 1997;43:690–2.
Method(s)	HPLC
Comment(s)	This study was performed on plasma obtained from healthy children. Results are the 2.5th and 97.5th percentiles.

HOMOVANILLIC ACID (HVA) (URINE) 4-HYDROXY-3-METHOXYPHENYLACETIC ACID

Male and Female							
Test	**Age**	**n**	**mg/g creatinine**	**mmol/mol creatinine**	**n**	**mg/24h**	**µmol/24h**
1	0–1 y	37	<32.6	<20.2	48	<2.8	<15.4
	2–4 y	49	<22.0	<13.6	34	<4.7	<25.8
	5–9 y	79	<15.1	<9.4	20	<5.4	<29.6
	10–19 y	55	<12.8	<7.9	40	<7.2	<39.5
	>19 y	56	<7.6	<4.7	56	<8.3	<45.6
2	0–3 mo	12	11.3–35.0	7.0–21.7			
	3–12 mo	30	8.4–44.9	5.2–27.8			
	1–2 y	16	12.2–31.8	7.6–19.7			
	2–5 y	21	3.4–32.0	2.1–19.8			
	5–10 y	25	6.8–23.7	4.2–14.7			
	10–15 y	13	3.2–13.6	2.0–8.4			
	>15 y	9	3.2–9.6	2.0–6.0			

Specimen Type(s)	1,2	Urine
Reference(s)	1	Soldin SJ, Hill JG. Liquid chromatographic analysis for urinary 4-hydroxy-3-methoxy-mandelic acid and 4-hydroxy-3-methoxy-phenylacetic acid and its use in investigation of neural crest tumors. Clin Chem 1981;27:502–3.
	2.	Tuchman M, Morris CL, Ramnaraine ML, et al. Value of random urinary homovanillic acid and vanillyl mandelic acid levels in the diagnosis and management of patients with neuroblastoma: comparison with 24-h urine collections. Pediatrics 1985;75: 324–8.
Method(s)	1	HPLC with electrochemical detection.
	2	Capillary gas chromatography.
Comment(s)	1	Analysis performed on patients under investigation of hypertension or not suspected of having a neural crest tumor. All patients studied free of neoplasia. Results are 95th percentile.
	2	Values are 0–100th percentiles. Normal values were obtained from 93 pediatric patients in whom the diagnosis of neural crest tumors had been excluded.

HYALURONIC ACID

Male and Female		
Age	**n**	**µg/L**
1–3 mo	*	49–153
2–3 y		9–40
4–18 y		6–32

Specimen Type(s)	Serum
Reference(s)	Trivedi P, Cheeseman P, Mowat AP. Serum hyaluronic acid in healthy infants and children and its value as a marker of progressive hepatobiliary disease starting in infancy. Clin Chim Acta 1993;215:29–39.
Method(s)	Affinity-binding radiometric assay (Pharmacia Diagnostics AB, Uppsala, Sweden).
Comment(s)	397 healthy infants. *See reference for details.

β–HYDROXYBUTYRATE/ 3–HYDROXYBUTYRATE

Male and Female			
Age	**n**	**mg/dL**	**mmol/L**
1–12 mo	12	1.0–10.0	.1–1.0
1–7 y	27	<1.0–9.4	<0.1–0.9
7–15 y	9	<1.0–3.1	<0.1–0.3

Specimen Type(s)	Whole blood precipitated with perchloric acid.
Reference(s)	Bonnefont JP, Specola NB, Vassault A, et al. The fasting test in paediatrics: application to the diagnosis of pathological hypo- and hyperketotic states. Eur J Pediatr; 1990;150:80–5.
Method(s)	Standard enzymatic procedure. See reference.
Comment(s)	The above results are 10–90th percentiles and refer to 15-h fasting values. For 20- and 24-h fasting values see reference.

5–HYDROXYINDOLEACETIC ACID (5HIAA) (URINE)

Male and Female				
Age	**n**	**mg/24h**	**µmol/24h**	**mg/g creatinine**
3–8 y	76	0.4–5.6	2.1–29.3	1.2–16.2
9–12 y	43	1.0–6.3	5.2–32.9	2.4–8.7
13–17 y	43	0.9–6.5	4.7–34.0	1.8–5.5
Adults	51	1.0–7.0	5.2–36.6	1.3–6.9

Specimen Type(s)	Urine
Reference(s)	Pediatric endocrine testing. Nichols Institute, 1993:24.
Method(s)	Fluorescent polarization.
Comment(s)	Results are 2.5–97.5th percentiles.

17α-HYDROXYPROGESTERONE (17α-OHP)

		Male			Female		
Test	Age	n	ng/dL	nmol/L	n	ng/dL	nmol/L
1	1–30 d	45	53–186	1.6–5.6	51	17–204	0.5–6.2
	31–182 d	48	35–157	1.1–4.8	46	25–110	0.8–3.3
	183–365 d	27	6–40	0.2–1.2	34	5–47	0.2–1.4
	1–3 y	104	2–19	0.1–0.6	77	3–51	0.1–1.5
	4–6 y	91	1–34	0.0–1.0	64	4–34	0.1–1.0
	7–9 y	74	1–45	0.0–1.4	53	4–44	0.1–1.3
	10–12 y	89	1–34	0.0–1.0	45	3–33	0.1–1.0
	13–15 y	61	23–82	0.7–2.5	42	2–72	0.1–2.2
	16–18 y	33	8–100	0.2–3.0	38	3–91	0.1–2.8

		Male and Female		
Test	Age	n	ng/dL	nmol/L
2	0–1 d	*	170–2500	5–75
	1–7 d		30–350	1–10
	7–28 d		0–250	0–8
	1–12 mo		0–170	0–5
	1–<4 y		0–100	0–3
	4–<10 y (P_1)		0–100	0–3
	P_2		0–130	0–4
	P_3		30–200	1–6
	P_4		30–230	1–7
	P_5		30–250	1–8

		Male and Female		
Test	Age	n	ng/dL	nmol/L
3	Premature infants	**		
	26–28 wks, day 4		124–841	3.7–25.2
	31–35 wks, day 4		26–568	0.8–17.0
	Full-term infants			
	3 d		7–77	0.2–2.3
	1–12 mo	**	13–106[a]	0.4–3.23[a]
	Prepubertal children			
	1–10 y	**	3–90	0.1–2.7

Test	Male			Female[b]		
	Age	ng/dL	nmol/L	Age	ng/dL	nmol/L
3	Puberty[c]			Puberty[c]		
cont	Tanner Stage[d]			Tanner Stage[d]		
	1 (<9.8 y)	3–90	0.1–2.7	1 (<9.2 y)	3–82	0.1–2.5
	2 (9.8–14.5 y)	5–115	0.2–3.5	2 (9.2–13.7 y)	11–98	0.3–2.9
	3 (10.7–15.4 y)	10–138	0.3–4.1	3 (10.0–14.4 y)	11–155	0.3–4.7
	4 (11.8–16.2 y)	29–180	0.9–5.4	4 (10.7–15.6 y)	18–230	0.5–6.9
	5 (12.8–17.3 y)	24–175	0.7–5.3	5 (11.8–18.6 y)	20–265	0.6–8.0
	n**			n**		

Specimen Type(s)	1,2	Plasma/serum
	3	Serum
Reference(s)	1	Soldin SJ, Bailey J, Beatey J, et al. Pediatric reference ranges for 17α-hydroxy-progesterone. Clin Chem 1995;41:S92. (Abstract)
	2	Soldin SJ, Rifai N, Hicks JM, eds. Biochemical basis of pediatric disease, 3rd ed. Washington, DC: AACC Press, 1998:233.
	3	Endocrinology expected values. Esoterix Endocrinology, Calabasas Hills, CA. ©2002 Esoterix, Inc. http://www.esoterix.com
		[c]Data adapted from Bidlingmaier F, Wagner-Barnack M, Butenandt O, Knorr D. Plasma estrogens in childhood and puberty under physiologic and pathologic conditions. Pediat Res 1973;7:901.
		[d]Tanner JM, Whitehouse RH. Clinical longitudinal standards for height, weight, height velocity, and the stages of puberty. Arch Dis Childhood 1976;51:170.
Method(s)	1	Coat-A-count (Diagnostic Products Corporation, Los Angeles, CA).
	2	See reference.
	3	Not provided.
Comment(s)	1	Study used hospitalized patients and a computerized approach adapted from the Hoffmann technique. Values are 2.5–97.5th percentiles.
	2	*These reference ranges are for guidance only. Numbers not provided. P_1–P_5 refers to pubertal stages.
	3	**Numbers not provided.
		[a]Levels increase after the first week to peak values ranging from 40–200 ng/dL between 30 and 60 days. Values then decline to prepubertal range before one year.
		[b]Days of cycle not determined for prepubertal females.
		[d]As used herein, Tanner stages in males encompass development of both pubic hair and genitalia. In females, each stage encompasses both pubic hair and breast development. While this expands the chronological age range for each stage of pubertal development, it results in a better correlation with hormonal values.

IGF BINDING PROTEIN-3 (IGF BP-3)

Male and Female			
Test	**Age**	**n**	**mg/L**
1	2–23 mo	40	0.7–2.3
	2–7 y	36	0.9–4.1
	8–11 y	68	1.5–4.3
	12–18 y	207	2.2–4.2
	19–55 y	137	2.0–4.0
2	0–1 wk	32	0.42–1.39
	1–4 wk	16	0.77–2.09
	1–3 mo	20	0.87–2.54
	3–6 mo	18	0.98–2.64
	6–12 mo	25	1.07–2.76
	1–3 y	36	1.41–2.97
	3–5 y	22	1.52–3.32
	5–7 y	25	1.66–3.59
	7–9 y	47	1.82–3.80
	9–11 y	50	2.12–4.26
	11–13 y	77	2.22–4.89
	13–15 y	77	2.31–5.24
	15–17 y	42	2.33–4.95
	20–30 y	32	2.20–4.93

Specimen Type(s)	1,2	Serum
Reference(s)	1	Pediatric endocrine testing. Nichols Institute, 1993;26.
	2	Blum WF, Ranke MB, Kietzmann K, et al. A specific radio-immunoassay for the growth hormone (GH)-dependent somatomedin-binding protein: Its use for diagnosis of GH deficiency. J Clin Endocrinol Metab 1990;70:1292–8.
Method(s)	1,2	Radioimmunoassay.
Comment(s)	1	Results are 2.5–97.5th percentiles.
	2	Results are 5–95th percentiles.

IMMUNOGLOBULIN A (IgA)

	Male and Female			
Test	Age	n	mg/dL	g/L
1	0–12 mo	75	0–83	0.00–0.83
	1–3 y	52	20–100	0.20–1.00
	4–6 y	41	27–195	0.27–1.95
	7–9 y	55	34–305	0.34–3.05
	10–11 y	38	53–204	0.53–2.04
	12–13 y	38	58–359	0.58–3.59
	14–15 y	38	47–249	0.47–2.49
	16–19 y	74	61–348	0.61–3.48

		Male			Female		
Test	Age	n	mg/dL	g/L	n	mg/dL	g/L
2	1–30 d	61	1–17	0.01–0.17	56	1–16	0.01–0.16
	31–182 d	52	6–47	0.06–0.47	60	1–49	0.01–0.49
	183–365 d	37	8–89	0.08–0.89	23	13–75	0.13–0.75
	1–3 y	105	15–142	0.15–1.42	127	21–117	0.21–1.17
	4–6 y	138	50–192	0.50–1.92	135	39–170	0.39–1.70
	7–9 y	111	64–209	0.64–2.09	105	34–181	0.34–1.81
	10–12 y	104	51–223	0.51–2.23	115	61–198	0.61–1.98
	13–15 y	104	35–252	0.35–2.52	118	68–246	0.68–2.46
	16–18 y	112	74–261	0.74–2.61	100	75–267	0.75–2.67
3	1–30 d	*	0–11	0–0.11	*	0–10	0–0.10
	31–182 d		0–40	0–0.40		0–42	0–0.42
	183–365 d		1–82	0.01–0.82		6–68	0.06–0.68
	1–3 y		9–137	0.09–1.37		15–111	0.15–1.11
	4–6 y		44–187	0.44–1.87		33–166	0.33–1.66
	7–9 y		58–204	0.58–2.04		28–180	0.28–1.80
	10–12 y		46–218	0.46–2.18		55–193	0.55–1.93
	13–15 y		29–251	0.29–2.51		62–241	0.62–2.41
	16–18 y		68–259	0.68–2.59		69–262	0.69–2.62

Specimen Type(s)	1	Plasma
	2,3	Plasma/serum
Reference(s)	The above values have been adjusted from those published in references 1 and 2 to convert results to current IFCC units using the new IFCC guidelines/standards.	
	1	Lockitch G, Halstead AC, Quigley G, et al. Age- and sex-specific pediatric reference intervals: Study design and methods illustrated by measurement of serum proteins with the Behring LN nephelometer. Clin Chem 1988;34:1618–21.
	2	Soldin SJ, Bailey J, Beatey J, et al. Pediatric reference ranges for immunoglobulins G, A and M on the Behring nephelometer. Clin Chem 1996;42:S308. (Abstract)
	3	Ghoshal AK, Soldin SJ. Evaluation of the Dade Behring Dimension RxL: integrated chemistry system-pediatric reference ranges. Clin Chim Acta 2003;331:135–46.
Method(s)	1,2	Nephelometry (Behring Diagnostics, Inc., Westwood, MA).
	3	The IgA method is a quantitative turbidimetric assay using endpoint detection based on the precipitation of IgA by its polyclonal antibody. Dade Behring Dimension RxL Analyzer (Dade Behring Inc., Newark, DE).
Comment(s)	1	Study performed on healthy normal children. Results are 2.5–97.5th percentiles.
	2	Study used hospitalized patients and a computerized approach adapted from the Hoffmann technique to obtain the 2.5–97.5th percentiles.
	3	*Numbers not provided.
		Reference ranges were obtained by comparing results from previously published data and using regression equations.

IMMUNOGLOBULIN D (IgD)

Male and Female				
Age	**n**	**mg/dL**	**n**	**mg/L**
Newborn	*	None detected	*	None detected
Thereafter		0–8		0–80

Specimen Type(s)	Serum
Reference(s)	Behrman RE, ed. Nelson textbook of pediatrics, 14th ed. Philadelphia, PA: WB Saunders Company, 1992:1813.
Method(s)	Not given.
Comment(s)	*Numbers not provided in reference.

IMMUNOGLOBULIN E (IgE)

		Male		Female	
Test	**Age**	**n**	**KIU/L**	**n**	**KIU/L**
1	0–12 mo	40	2–24	29	0–20
	1–3 y	74	2–149	104	2–55
	4–10 y	155	4–249	124	8–279
	11–15 y	137	7–280	152	5–295
	16–18 y	58	5–268	67	7–698
2	0–12 mo	28	<12	28	<8
	1–3 y	44	<90	67	<28
	4–10 y	76	<163	56	<137
	11–18 y	96	<179	109	<398
3	0–12 mo	*	<20	*	<17
	1–3 y		<85		<33
	4–10 y		<146		<124
	11–18 y		<159		<177

Specimen Type(s)	1–3	Plasma/serum
Reference(s)	1	Soldin SJ, Morales A, Albalos F, et al. Pediatric reference ranges on the Abbott Imx for FSH, LH, prolactin, TSH, T_4, T_3, free T_4, free T_3, T-uptake, IgE and ferritin. Clin Biochem 1995;28:603–6.
	2	Soldin SJ, Lenherr S, Kumar A. Pediatric reference ranges for IgE. Clin Chem 1995;41:S92. (Abstract)
	3	DPC IMMULITE® 1000 reference ranges used at Children's National Medical Center, Washington, DC.
Method(s)	1	Abbott Imx (Abbott Laboratories, Abbott Park, IL).
	2	Behring LN nephelometer (Behring Diagnostics, Westwood, MA).
	3	DPC IMMULITE® 1000 analyzer (Diagnostic Products Corporation, Los Angeles, CA).
Comment(s)	1,2	Study performed on hospitalized patients and the Hoffmann technique applied to obtain the 2.5–97.5th percentiles (1) and 97.5th percentile (2).
	3	*Numbers not provided.
		Pediatric reference ranges were determined by using regression equations to transform the previously established ranges on the Bayer Immuno 1 (Bayer Corp., Tarrytown, NY) based on the Hoffmann technique to their corresponding values on the IMMULITE® 1000.

IMMUNOGLOBULIN G (IgG)

Male and Female				
Test	Age	n	mg/dL	g/L
1	0–12 mo	75	232–1411	2.32–14.11
	1–3 y	52	453–916	4.53–9.16
	4–6 y	41	504–1465	5.04–14.65
	7–9 y	55	572–1474	5.72–14.74
	10–11 y	38	698–1560	6.98–15.60
	12–13 y	38	759–1550	7.59–15.50
	14–15 y	38	716–1711	7.16–17.11
	16–19 y	74	549–1584	5.49–15.84

		Male			Female		
Test	Age	n	mg/dL	g/L	n	mg/dL	g/L
2	1–30 d	61	221–838	2.21–8.38	56	188–876	1.88–8.76
	31–182 d	52	166–547	1.66–5.47	60	332–675	3.32–6.75
	183–365 d	37	156–829	1.56–8.29	23	346–658	3.46–6.58
	1–3 y	105	431–1109	4.31–11.09	127	468–1196	4.68–11.96
	4–6 y	138	485–1318	4.851–13.18	135	574–1309	5.74–13.09
	7–9 y	111	595–1428	5.95–14.28	105	501–1459	5.01–14.59
	10–12 y	104	695–1602	6.95–16.02	115	599–1590	5.99–15.90
	13–15 y	104	603–1582	6.03–15.82	118	757–1621	7.57–16.21
	16–18 y	112	537–1682	5.37–16.82	100	810–1792	8.10–17.92
3	1–30 d	*	197–833	1.97–8.33	*	162–872	1.62–8.72
	31–182 d		140–533	1.40–5.33		311–664	3.11–5.33
	183–365 d		130–156	1.30–1.56		325–647	3.25–6.47
	1–3 y		413–1112	4.13–11.12		451–1202	4.51–12.02
	4–6 y		468–1328	4.68–13.28		560–1319	5.60–13.19
	7–9 y		582–1441	5.82–14.41		485–1473	4.85–14.73
	10–12 y		685–1620	6.85–16.20		586–1609	5.86–16.09
	13–15 y		590–1600	5.90–16.00		749–1640	7.49–16.40
	16–18 y		522–1703	5.22–17.03		804–1817	8.04–18.17

Specimen Type(s)	1	Plasma
	2,3	Plasma/serum
Reference(s)		The above values have been adjusted from those published in references 1 and 2 to convert results to current IFCC units using the new IFCC guidelines/standards.
	1	Lockitch G, Halstead AC, Quigley G, et al. Age- and sex-specific pediatric reference intervals: Study design and methods illustrated by measurement of serum proteins with the Behring LN nephelometer. Clin Chem 1988;34:1618–21.
	2	Soldin SJ, Bailey J, Beatey J, et al. Pediatric reference ranges for immunoglobulins G, A and M on the Behring nephelometer. Clin Chem 1996;42:S308. (Abstract)
	3	Ghoshal AK, Soldin SJ. Evaluation of the Dade Behring Dimension RxL: integrated chemistry system-pediatric reference ranges. Clin Chim Acta 2003;331:135–46.
Method(s)	1,2	Nephelometry (Behring Diagnostics, Inc., Westwood, MA).
	3	The IgG method is a quantitative turbidimetric assay using endpoint detection based on the precipitation of IgG by its polyclonal antibody. Dade Behring Dimension RxL Analyzer (Dade Behring Inc., Newark, DE).
Comment(s)	1	Study performed on healthy normal children.
		Results are 2.5–97.5th percentiles.
	2	Study used hospitalized patients and a computerized approach adapted from the Hoffmann technique to obtain the 2.5–97.5th percentiles.
	3	*Numbers not provided.
		Reference ranges were obtained by comparing results from previously published data and using regression equations.

IMMUNOGLOBULIN M (IgM)

Test	Age	n	Male and Female mg/dL	Male and Female g/L
1	0–12 mo	75	0–145	0.00–1.45
	1–3 y	52	19–146	0.19–1.46
	4–6 y	41	24–210	0.24–2.10
	7–9 y	55	32–208	0.32–2.08
	10–11 y	38	31–180	0.31–1.80
	12–13 y	38	35–239	0.35–2.39
	14–15 y	38	15–188	0.15–1.88
	16–19 y	74	23–259	0.23–2.59

		Male			Female		
Test	Age	n	mg/dL	g/L	n	mg/dL	g/L
2	1–30 d	61	8–78	0.08–0.78	56	13–70	0.13–0.70
	31–182 d	52	18–98	0.18–0.98	60	6–142	0.06–1.42
	183–365 d	37	27–132	0.27–1.32	23	3–145	0.03–1.45
	1–3 y	105	42–161	0.42–1.61	127	47–200	0.47–2.00
	4–6 y	138	43–166	0.43–1.66	135	54–200	0.54–2.00
	7–9 y	111	33–155	0.33–1.55	105	42–181	0.42–1.81
	10–12 y	104	39–167	0.39–1.67	115	54–228	0.54–2.28
	13–15 y	104	38–200	0.38–2.00	118	46–242	0.46–2.42
	16–18 y	112	40–195	0.40–1.951	100	58–241	0.58–2.41
3	1–30 d	*	0–65	0–0.65	*	1–57	0–0.57
	31–182 d		6–84	0.06–0.84		0–127	0–1.27
	183–365 d		15–117	0.15–1.17		0–130	0–1.30
	1–3y		30–146	0.30–1.46		35–184	0.35–1.84
	4–6 y		31–151	0.31–1.51		42–184	0.42–1.84
	7–9 y		21–140	0.21–1.40		30–165	0.30–1.65
	10–12 y		27–151	0.27–1.51		42–211	0.42–2.11
	13–15 y		26–184	0.26–1.84		34–225	0.34–2.25
	16–18 y		28–179	0.28–1.79		45–224	0.45–2.24

Specimen Type(s)	1	Plasma
	2,3	Plasma/serum
Reference(s)		The above values have been adjusted from those published in references 1 and 2 to convert results to current IFCC units using the new IFCC guidelines/standards.
	1	Lockitch G, Halstead AC, Quigley G, et al. Age- and sex-specific pediatric reference intervals: Study design and methods illustrated by measurement of serum proteins with the Behring LN nephelometer. Clin Chem 1988;34:1618–21.
	2	Soldin SJ, Bailey J, Beatey J, et al. Pediatric reference ranges for immunoglobulins G, A and M on the Behring nephelometer. Clin Chem 1996;42:S308. (Abstract)
	3	Ghoshal AK, Soldin SJ. Evaluation of the Dade Behring Dimension RxL: integrated chemistry system-pediatric reference ranges. Clin Chim Acta 2003;331:135–46.
Method(s)	1,2	Nephelometry (Behring Diagnostics, Inc., Westwood, MA).
	3	The IgM method is a quantitative turbidimetric assay using endpoint detection based on the precipitation of IgM by its polyclonal antibody. Dade Behring Dimension RxL Analyzer (Dade Behring Inc., Newark, DE).
Comment(s)	1	Study performed on healthy normal children. Results are 2.5–97.5th percentiles.
	2	Study used hospitalized patients and a computerized approach adapted from the Hoffmann technique to obtain the 2.5–97.5th percentiles.
	3	*Numbers not provided.
		Reference ranges were obtained by comparing results from previously published data and using regression equations.

IMMUNOGLOBULIN LIGHT CHAINS (KAPPA AND LAMBDA)

		Male and Female			
		Kappa		Lambda	
Age	n	mg/dL	g/L	mg/dL	g/L
Newborn	325	770–870	7.7–8.7	170–190	1.7–1.9
Premature	168	310–490	3.1–4.9	160–190	1.6–1.9
1 mo	50	360–480	3.6–4.8	160–200	1.6–2.0
2 mo	50	240–270	2.4–2.7	160–170	1.6–1.7
3 mo	50	170–250	1.7–2.5	100–130	1.0–1.3
4 mo	50	190–240	1.9–2.4	80–100	0.8–1.0
5 mo	50	200–320	2.0–3.2	100–110	1.0–1.1
1 y	50	300–530	3.0–5.3	90–120	0.9–1.2
2 y	50	360–640	3.6–6.4	130–150	1.3–1.5
3 y	50	410–540	4.1–5.4	110–140	1.1–1.4
4 y	50	480–830	4.8–8.3	120–170	1.2–1.7
5 y	50	540–710	5.4–7.1	130–170	1.3–1.7
6 y	50	600–850	6.0–8.5	160–170	1.6–1.7
7 y	50	460–750	4.6–7.5	140–180	1.4–1.8
8 y	50	730–910	7.3–9.1	140–170	1.4–1.7
9 y	50	710–960	7.1–9.6	150–180	1.5–1.8
10 y	50	820–960	8.2–9.6	130–190	1.3–1.9
11 y	50	600–830	6.0–8.3	140–180	1.4–1.8
12 y	50	820–960	8.2–9.6	170–200	1.7–2.0
13 y	50	820–1050	8.2–10.5	140–170	1.4–1.7
14 y	50	800–1080	8.0–10.8	190–220	1.9–2.2
15 y	50	720–1100	7.2–11.0	170–230	1.7–2.3
16 y	50	700–1120	7.0–11.2	150–220	1.5–2.2

Specimen Type(s)	Serum
Reference(s)	Herkner KR, Salzer H, Böck A, et al. Pediatric and perinatal reference intervals for immunoglobulin light chains kappa and lambda. Clin Chem 1992;38:548–50.
Method(s)	Nephelometry. Kallestad Model CPM 300 (Diagnostics Pasteur, Marnes-la-Coquette, France).
Comment(s)	The subjects were healthy newborns and children who had been screened to rule out immunological disorders and infections. The ranges are 10–90th percentiles.

IMMUNOGLOBULIN G SUBCLASS (IgG-SUBCLASS)

Male and Female, g/L (To convert to mg/dL, multiply by 100)					
Age	**n**	**IgG_1**	**IgG_2**	**IgG_3**	**IgG_4**
Cord blood serum preterm	20	3.4–9.7	0.7–1.7	0.2–0.5	0.2–0.7
Cord blood serum term	20	5.8–13.7	0.6–5.2	0.2–1.2	0.2–1.0
5 y	20	5.6–12.7	0.4–4.4	0.3–1.0	0.1–0.8
6 y	20	6.2–11.3	0.5–4.0	0.3–0.8	0.2–0.9
7 y	20	5.4–10.5	0.9–3.5	0.3–1.1	0.2–1.1
8 y	20	5.6–10.5	0.7–4.5	0.2–1.1	0.1–0.8
9 y	20	3.9–11.4	0.7–4.7	0.4–1.2	0.2–1.0
10 y	20	4.4–10.8	0.6–4.0	0.3–1.2	0.1–0.9
11 y	20	6.4–10.9	0.9–4.3	0.3–0.9	0.2–1.0
12 y	20	6.0–11.5	0.9–4.8	0.4–1.0	0.2–0.9
13 y	20	6.1–11.5	0.9–7.9	0.2–1.1	0.1–0.8

Specimen Type(s)	Serum
Reference(s)	Miles J, Riches P. The determination of IgG subclass concentrations in serum by enzyme linked immunosorbent assay: establishment of age-related reference ranges for cord blood samples, children aged 5–13 years and adults. Ann Clin Biochem 1994;31:24–8.
Method(s)	ELISA
Comment(s)	Cord blood samples were obtained from both premature and term infants. Samples were obtained from school children and adults. Results are 5–95th percentiles.

INSULIN-LIKE GROWTH FACTOR-1 (IGF-1), SOMATOMEDIN-C

		Male			Female		
Test	**Age**	**n**	**nmol/L**	**ng/mL**	**n**	**nmol/L**	**ng/mL**
1	1–30 d	34	0.2–7.3	2–56	31	0.9–11.9	7–92
	31–182 d	55	0.2–10.6	2–82	61	0.6–9.4	5–72
	183–365 d	27	0.2–7.0	2–54	35	1.1–9.8	8–75
	1–3 y	140	2.2–15.0	17–116	111	1.6–17.5	12–135
	4–6 y	126	2.7–20.8	21–160	83	1.7–22.8	13–176
	7–9 y	100	8.4–26.9	65–207	74	6.9–32.9	53–253
	10–12 y	113	9.0–25.4	69–196	86	9.8–46.3	75–357
	13–15 y	116	10.4–52.5	80–404	94	8.6–44.7	66–344
	16–18 y	62	12.5–49.7	96–383	85	12.8–47.3	99–364
2	2 mo–5 y	97	2.2–32.2	17–248	97	2.2–32.2	17–248
	6–8 y	40	11.4–61.5	88–474	40	11.4–61.5	88–474
	9–11 y	125	14.2–73.4	110–565	174	15.2–100.1	117–771
	12–15 y	135	26.2–124.3	202–957	195	33.9–142.3	261–1096
	16–24 y	95	23.6–101.3	182–780	95	23.6–101.3	182–780

Specimen Type(s)	1,2	Serum
Reference(s)	1	Soldin SJ, Hicks JM, Bailey J, et al. Pediatric reference ranges for creatine kinase and insulin-like growth factor 1. Clin Chem 1997;43:S199. (Abstract)
	2	Nichols Institute Test Catalogue, 1996.
Method(s)	1	Radioimmunoassay, IncStar® (IncStar Corporation, Stillwater, MN).
	2	Radioimmunoassay.
Comment(s)	1	Results are 2.5–95th percentiles. Study used hospitalized patients and a computerized adaptation of the Hoffmann technique.
	2	Results are 2.5–97.5th percentiles.

INSULIN-LIKE GROWTH FACTOR-II (IGF-II)

		Males		Females	
Test	Age	n	ng/mL µg/L	n	ng/mL µg/L
1	<3 y	10	310–650	10	129–709
	3–5 y	9	55–1091	15	124–928
	6–8 y	15	237–977	20	145–1057
	9–11 y	10	127–883	16	252–792
	12–14 y	38	272–728	26	266–730
	>15 y	14	317–745	14	176–756

Specimen Type(s)	Plasma
Reference(s)	Rosenfeld RG, Wilson DM, Lee PDK, et al. Insulin-like growth factors I and II in evaluation of growth retardation. J Ped 1986;109:428–33.
Method(s)	RIA after gel filtration on Sephadex G-50. See reference above.
Comment(s)	Subjects were normal healthy children, with heights between the 5th and 95th percentiles for age, selected from an outpatient pediatric clinic. Results are 2.5–97.5th percentiles.

IODIDE (URINE), (UI)

Age	Male			Female		
	n	UI Concentration µg/dL	UI/Creatinine Ratio (UI/CR) µg/g	n	UI Concentration µg/dL	UI/Creatinine Ratio (UI/CR) µg/g
6–9 y	1025	5.4–85.0	80–940	950	3.6–74.0	68–947
10–11 y	553	6.0–98.0	62–820	520	4.1–80.0	57–744
12–13 y	390	4.5–74.0	51–840	429	3.6–75.0	42–564
14–44 y	4782	2.4–66.0	34–420	5522	1.8–65.0	36–539

Female			
Age	n	UI Concentration µg/dL	UI/Creatinine Ratio (UI/CR) µg/g
Child-bearing age (14–44 y)			
Pregnant	290	4.2–55.0	33–535
Not Pregnant	5232	1.8–65.0	36–541
All Females	5522	1.8–65.0	36–539

Specimen Type(s)	Urine
Reference(s)	Soldin OP, Soldin SJ, Pezzullo JC. Urinary iodine percentile ranges in the United States. Clin Chim Acta, 2003;328:185–90.
Method(s)	Sandell-Koltoff reaction. Benotti J, Benotti N, Pino S, et al. Determination of the total iodine in urine, stool, diets, and tissue. Clin Chem 1965;11:932–6.
Comment(s)	The third National Health and Nutrition Examination Survey (NHANES III) (1998–1994) database of the civilian, non-institutionalized, iodide-sufficient US population was used. Gunter EW, Lewis BL, Konchikowski SM. Laboratory methods used for the Third National Health and Nutrition Examination Survey (NHANES III), 1988–1994. Hyattsville: Centers for Disease Control and Prevention, 1996.

IRON

		Male			Female		
Test	**Age**	**n**	**µg/dL**	**µmol/L**	**n**	**µg/dL**	**µmol/L**
1	1–5 y*	44*	22–136	4–25	44*	22–136	4–25
	6–9 y*	50*	39–136	7–25	50*	39–136	7–25
	10–14 y	31	28–134	5–24	40	45–145	8–26
	14–19 y	65	34–162	6–29	110	28–184	5–33
2	1–30 d	80	32–112	5.7–20.0	44	29–127	5.2–22.7
	31–365 d	87	27–109	4.8–19.5	75	25–126	4.5–22.6
	1–3 y	160	29–91	5.2–16.3	119	25–101	4.5–18.1
	4–6 y	114	25–115	4.5–20.6	88	28–93	5.0–16.7
	7–9 y	116	27–96	4.8–17.2	101	30–104	5.4–18.6
	10–12 y	107	28–112	5.0–20.0	91	32–104	5.7–18.6
	13–15 y	98	26–110	4.7–19.7	117	30–109	5.4–19.5
	16–18 y	92	27–138	4.8–24.7	99	33–102	5.9–18.3
Male and Female							
		5–11 am			**5–11 pm**		
Test	**Age**	**n**	**µg/dL**	**µmol/L**	**n**	**µg/dL**	**µmol/L**
3,4	0–24 mo	332	20–105	3.6–18.8	43	20–140	3.6–25.1
	2–9 y	370	20–105	3.6–18.8	100	20–145	3.6–26.0
	10–14 y	145	20–100	3.6–17.9	43	20–145	3.6–26.0
	15–18 y	125	20–100	3.6–17.9	42	20–145	3.6–26.0

Specimen Type(s)	1	Plasma
	2–4	Plasma/serum
Reference(s)	1	Lockitch G, Halstead A, Wadsworth L, et al. Age- and sex-specific pediatric reference intervals for zinc, copper, selenium, iron, vitamins A and E, and related proteins. Clin Chem 1988;34:1625–8.
	2	Soldin SJ, Bailey J, Bjorn J, et al. Pediatric reference ranges for iron on the Hitachi 747 with Boehringer Mannheim reagents. Clin Chem 1999;45:A22. (Abstract)
	3	Soldin SJ, Murthy JN, Agarwalla Pk, et al. Pediatric reference ranges for creatine kinase, CKMB, troponin I, iron and cortisol. Clin Biochem 1999;32:77–80.
	4	Ghoshal AK, Soldin SJ. Evaluation of the Dade Behring Dimension RxL: integrated chemistry system-pediatric reference ranges. Clin Chim Acta 2003;331:135–46.
Method(s)	1	Ferrozine (Sigma Chemical Company). RA-100 Analyzer (Technicon Instruments, Tarrytown, NY).
	2	Hitachi 747 with Boehringer Mannheim reagents (Boehringer Mannheim Diagnostics, Indianapolis, IN).
	3	Vitros 500 analyzer using Vitros reagents (Ortho-Clinical Diagnostics, Raritan, NJ).
	4	Under acidic conditions (pH 4.5), iron bound to the protein transferrin is released in the presence of the reducing agent, ascorbic acid. The resulting product, Fe^{++}, forms a blue complex with 3-(2-pyridyl)-5,6-bis-2-(5-furyl sulfonic acid)-1,2,4-triazine, disodium salt (Ferene®). Dade Behring Dimension RxL Analyzer (Dade Behring Inc., Newark, DE).
Comment(s)	1	The study population was healthy children. Non-parametric methods were used to determine the 0.025 and 0.975 fractiles. *No significant differences were found for males and females. These ranges were therefore derived from combined data.
	2–4	Study used hospitalized patients and a computerized approach adapted from the Hoffmann technique. Values are the 2.5–97.5th percentiles. Results are somewhat lower than expected and represent data from a largely inner city population.

IRON BINDING CAPACITY, TOTAL (TIBC)

		Male			Female		
Test	**Age**	**n**	**µg/dL**	**µmol/L**	**n**	**µg/dL**	**µmol/L**
1	1–5 y*	44	268–441	48–79	44	268–441	48–79
	6–9 y*	50	240–508	43–91	50	240–508	43–91
	10–14 y	31	302–508	54–91	40	318–575	57–103
	14–19 y	65	290–570	52–102	110	302–564	52–101
2	1–30 d	133	94–232	16.8–41.5	57	94–236	16.8–42.2
	31–182 d	78	116–322	20.8–57.6	69	89–311	15.9–55.7
	183–365 d	39	176–384	31.5–68.7	27	138–365	24.7–65.3
	1–3 y	131	204–382	36.5–68.3	103	184–377	32.9–67.5
	4–6 y	114	180–390	32.2–69.8	107	162–352	29.0–63.0
	7–9 y	116	183–369	32.8–66.1	113	167–336	29.9–60.1
	10–12 y	104	173–356	31.0–63.7	112	198–383	35.4–68.6
	13–15 y	106	193–377	34.5–67.5	143	169–358	30.3–64.1
	16–18 y	113	174–351	31.1–62.8	137	194–372	34.7–66.6

Specimen Type(s)	1	Plasma
	2	Serum
Reference(s)	1	Lockitch G, Halstead A, Wadsworth L, et al. Age- and sex-specific pediatric reference intervals for zinc, copper, selenium, iron, vitamins A and E, and related proteins. Clin Chem 1988;34:1625–8.
	2	Soldin SJ, Hicks JM, Bailey J, et al. Pediatric reference ranges for total iron binding capacity and transferrin. Clin Chem 1997;43:S200. (Abstract)
Method(s)	1	Transferrin was measured by nephelometry. Behring LN with Behring kit (Behringwerke, Marburg, Germany). Calculated TIBC.
	2	Hitachi 747 using Boehringer Mannheim reagents (Boehringer Mannheim Diagnostics, Indianapolis, IN).
Comment(s)	1	Values are 2.5–97.5th percentiles. Transferrin iron-binding capacity calculated by multiplying the transferrin g/L × 23.1. *No significant differences were found for males and females. These ranges were therefore derived from combined data.
	2	Study used hospitalized patients and a computerized approach adapted from the Hoffmann technique. Values are 2.5–97.5th percentiles.

LACTATE

Male and Female				
Test	**Age**	**n**	**mg/dL**	**mmol/L**
1	1–12 mo	12	10–21	1.1–2.3
	1–7 y	27	7–14	0.8–1.5
	7–15 y	9	5–8	0.6–0.9
2	0–90 d	85	30	3.3
	3–24 mo	150	28	3.1
	2–18 y	185	20	2.2
3	0–90 d	*	9–32	1.0–3.5
	3 mo–24 mo		9–30	1.0–3.3
	2–18 y		9–22	1.0–2.4

Specimen Type(s)	1	Whole blood precipitated with perchloric acid.
	2	Plasma
	3	Plasma/serum
Reference(s)	1	Bonnefont JP, Specola NB, Vassault A, et al. The fasting test in pediatrics: application to pathological hypo- and hyperketotic states. Eur J Pediatr 1990;150:80–5.
	2	Soldin SJ, Baumel CR. Pediatric reference ranges for CSF protein and plasma lactate on the Vitros 500 analyzer. Clin Chem 2001:47:A108.
	3	Ghoshal AK, Soldin SJ. Evaluation of the Dade Behring Dimension RxL: integrated chemistry system-pediatric reference ranges. Clin Chim Acta 2003;331:135–46.
Method(s)	1	Standard enzymatic procedure. See reference.
	2	Vitros 500 analyzer using Ortho-Clinical Diagnostics reagents (Ortho-Clinical Diagnostics, Raritan, NJ).
	3	The lactic acid method employs the oxidation of lactate to pyruvate. Dade Behring Dimension RxL Analyzer (Dade Behring Inc., Newark, DE).
Comment(s)	1	The above results are 10–90th percentiles and refer to 15-h fasting values. For 20- and 24-h fasting values, see reference.
	2	Studies used hospitalized patients and a computerized approach adapted from the Hoffmann technique. Values are 97.5th percentile.
	3	*Numbers not provided.
		Reference ranges were obtained by comparing results from previously published data and using regression equations.

LACTATE DEHYDROGENASE (LDH)

		Male		Female	
Test	Age	n	U/L	n	U/L
1	1–30 d	119	550–2100	76	580–2000
	1–3 mo	108	480–1220	84	460–1150
	4–6 mo	100	400–1230	54	480–1150
	7–12 mo	97	380–1200	75	460–1060
2	1–3 y	50*	500–920	50*	500–920
	4–6 y	40*	470–900	40*	470–900
	7–9 y	80*	420–750	80*	420–750
	10–11 y	27	432–700	34	380–700
	12–13 y	31	470–750	49	380–640
	14–15 y	26	360–730	52	390–580
	16–19 y	40	340–670	61	340–670
3	1–30 d	77	125–735	65	145–765
	31–365 d	86	170–450	74	190–420
	1–3 y	135	155–345	109	165–395
	4–6 y	101	155–345	96	135–345
	7–9 y	125	145–300	104	140–280
	10–12 y	111	120–325	76	120–260
	13–15 y	105	120–290	101	100–275
	16–18 y	79	105–235	98	105–230
4	1–30 d	**	178–629	**	187–600
	1–3 mo		158–373		152–353
	4–6 mo		135–376		158–353
	7–12 mo		129–367		152–327
	1–3 y		164–286		164–286
	4–6 y		155–280		155–280
	7–9 y		141–237		141–237
	10–11 y		141–231		129–222
	12–13 y		141–231		129–205
	16–19 y		117–217		117–213

Specimen Type(s)	1,4	Plasma/serum
	2,3	Plasma
Reference(s)	1	Soldin SJ, Savwoir TV, Guo Y. Pediatric reference ranges for lactate dehydrogenase and uric acid during the first year of life on the Vitros 500 analyzer. Clin Chem 1997;43:S199. (Abstract)
	2	Lockitch G, Halstead AC, Albersheim S, et al. Age- and sex-specific pediatric reference intervals for biochemistry analytes as measured with the Ektachem-700 analyzer. Clin Chem 1988;34:1622–5.
	3	Soldin SJ, Bailey J, Bjorn S, et al. Pediatric reference ranges for LDH. Clin Chem 1995;41:S93. (Abstract)
	4	Ghoshal AK, Soldin SJ. Evaluation of the Dade Behring Dimension RxL: integrated chemistry system-pediatric reference ranges. Clin Chim Acta 2003;331:135–46.
Method(s)	1,2	Vitros 500 (1) and 700 (2) (Ortho-Clinical Diagnostics, Raritan, NJ).
	3	Measured on the Hitachi 747 using Boehringer Mannheim reagents (Boehringer Mannheim Diagnostics, Indianapolis, IN).
	4	The LDH method measures the oxidation of L-lactate to pyruvate with simultaneous reduction of nicotinamide adenine dinucleotide. Dade Behring Dimension RxL Analyzer (Dade Behring Inc., Newark, DE).
Comment(s)	1,3	Study used hospitalized patients and a computerized approach adapted from the Hoffmann technique to obtain the 2.5–97.5th percentiles.
	2	The study population was healthy children. Non-parametric methods were used to determine the 0.025 and 0.975 fractiles.
		*No significant differences were found for males and females. These ranges were therefore derived from combined data.
	4	**Numbers not provided.
		Reference ranges were obtained by comparing results from previously published data and using regression equations.

LACTATE/PYRUVATE RATIO

	Male and Female		
Age	n	Fed	15-h Fast
1 mo–1 y	12	—	10–28
1–7 y	27	12–18	11–18
7–15 y	9	8–20	8–20

Specimen Type(s)	See lactate and pyruvate.
Reference(s)	Hommes FA, ed. Techniques in diagnostic human biochemical genetics. New York, NY: Wiley-Liss, 1991:300.
Method(s)	See references for lactate and pyruvate.
Comment(s)	Results are 10–90th percentiles.

LEAD

	Male		Female	
Age	n	µg/dL	n	µg/dL
0–12 mo	1072	1.8–4.9	710	2.3–4.7
13–24 mo	593	1.7–5.6	455	2.2–5.2
2–4 y	2513	1.6–5.5	3004	2.2–5.1
5–9 y	1053	1.6–5.2	1025	2.5–4.8
10–11 y	232	1.3–4.9	161	1.9–4.5
12–18 y	121	1.8–4.4	207	2.1–4.4

Specimen Type(s)	EDTA whole blood
Reference(s)	Soldin OP, Hanak B, Soldin SJ. Blood lead concentrations in children: new ranges. Clin Chim Acta 2003;327:109–13.
Method(s)	Electrothermal atomization atomic absorption spectrophotometric assay, AAnalyst 600 (Perkin Elmer, Wilton, CT).
Comment(s)	CDC guidelines place the threshold for concern at 10 µg/dL.

LIPASE

Male and Female			
Test	**Age**	**n**	**U/L**
1	1–30 d	83	6–55
	31–182 d	121	4–29
	183–365 d	56	4–23
	1–3 y	148	4–31
	4–9 y	96	3–32
	10–18 y	142	4–29

		Male		Female	
Test	**Age**	**n**	**U/L**	**n**	**U/L**
2	0–90 d	39*	10–85	39*	10–85
	3–12 mo	113	13–95	137	9–128
	1–<2 y	118	15–135	222	15–150
	2–<7 y	323	15–175	243	10–150
	7–<11 y	216	10–175	209	13–150
	11–<15 y	75	10–195	150	10–180
	15–18 y	70	10–195	79	10–220
3	0–90 d	**	145–174	**	145–174
	3–12 mo		146–178		144–190
	1–<2 y		147–193		147–199
	2–<7 y		147–209		145–199
	7–<11 y		145–209		146–199
	11–<15 y		145–216		145–211
	15–18 y		145–216		145–226

Specimen Type(s)	1–3	Plasma/serum
Reference(s)	1	Soldin SJ, Bailey J., Beatey J., et al. Pediatric reference ranges for lipase. Clin Chem 1995; 41:S93. (Abstract)
	2	Soldin SJ, Ojeifo O. Pediatric reference ranges for lipase. Clin Chem 1999;45:A22. (Abstract)
	3	Ghoshal AK, Soldin SJ. Evaluation of the Dade Behring Dimension RxL: integrated chemistry system-pediatric reference ranges. Clin Chim Acta 2003;331:135–46.
Method(s)	1	Hitachi 717 using Sigma reagents (Boehringer Mannheim Diagnostics, Indianapolis, IN).
	2	Vitros 500 using Vitros reagents (Ortho-Clinical Diagnostics, Raritan, NJ).
	3	The LIP method technology uses a glycerol derivative esterified in the 1-position with a dicarbonic acid-resorufin ester as the substrate. Lipase activity is maximized in the presence of colipase, calcium chloride, and bile salt. Dade Behring Dimension RxL Analyzer (Dade Behring Inc., Newark, DE).
Comment(s)	1,2	Study used hospitalized patients and a computerized approach to removing outliers. Values are 2.5–97.5th percentiles.
		*Male and female date combined.
	3	**Numbers not provided.
		Reference ranges were obtained by comparing results from previously published data and using regression equations.

LOW-DENSITY LIPOPROTEIN-CHOLESTEROL (LDL-C)

Male and Female				
Test	**Age**	**n**	**mmol/L**	**mg/dL**
1	2–12 mo	100	0.82–3.02	32–117
	2–10 y	120	0.98–3.62	38–140

		Male			Female		
Test	**Age**	**n**	**mmol/L**	**mg/dL**	**n**	**mmol/L**	**mg/dL**
2	5–9 y	*	1.63–3.34	63–129	*	1.76–3.63	68–140
	10–14 y		1.66–3.44	64–133		1.76–3.52	68–136
	15–19 y		1.61–3.37	62–130		1.53–3.55	59–137

Specimen Type(s)	1,2	Serum
Reference(s)	1	Baroni S, Scribano D, Valentini P, et al. Serum apolipoprotein A1, B, CII, CIII, E, and lipoprotein (a) levels in children. Clin Biochem 1996;29:603–5.
	2	Soldin SJ, Rifai N, Hicks JM, eds. Biochemical basis of pediatric disease, 3rd ed. Washington, DC: AACC Press, 1998:468.
Method(s)	1	Calculated as described in Friedewald WT, Levy RI, Frederickson DC. Estimation of the concentration of low-density lipoprotein cholesterol in plasma, without use of the preparative ultracentrifuge. Clin Chem 1972;18:499–502.
	2	See references.
Comment(s)	1	Results are the 2.5–97.5th percentiles for healthy children.
	2	*Numbers of participants not provided. Results are 5–95th percentiles.

LUTEINIZING HORMONE (LH)

		Male		Female	
Test	**Age**	**n**	**U/L**	**n**	**U/L**
1	<2 y	179	0.5–1.9	33	<0.5
	2–5 y	138	<0.5	94	<0.5
	6–10 y	96	<0.5	136	<0.5
	11–20 y	89	0.5–5.3	263	0.5–9.0
2	<2 y	179	1.3–2.8	33	<1.3
	2–5 y	138	<1.3	94	<1.3
	6–10 y	96	<1.3	136	<1.3
	11–20 y	89	1.3–5.7	263	1.3–10.2
3	0–1 y	*	1.7–2.7	*	0.1–1.7
	2–5 y		0.1–1.7		0.1–1.7
	6–10 y		0.1–1.7		0.1–1.7
	11–18 y		1.7–4.7		1.7–7.7

			Male	Female
Test	**Age**	**n**	**mIU/mL**	**mIU/mL**
4	0–7 d	**	1.5–5.0	1.5–5.0
	2 wk–1 y		Increases by 2 weeks to 3.0–22.0. Declines to prepubertal levels by 1 year.	Increases by 2 weeks to 1.8–13.0. Declines to prepubertal levels by 1 year.
	Prepubertal children		1.0–3.5	1.0–3.5
	Pubertal children		Rise to adult values: 3.0–10.0	Rise to adult values: Follicular 3.0–11.0 Mid-cycle 18.0–70.0 Luteal 1.0–11.0

Specimen Type(s)	1–3	Plasma/serum
	4	Serum
Reference(s)	1	Soldin SJ, Morales A, Albalos F, et al. Pediatric reference ranges on the Abbott IMx for FSH, LH, prolactin, TSH, T_4, T_3, free T_4, free T_3, T-uptake, IgE and ferritin. Clin Biochem 1995;28:603–6.
	2	Murthy JN, Hicks JM, Soldin SJ. Evaluation of the Technicon Immuno I Random Access Immunoassay Analyzer and calculation of pediatric reference ranges for endocrine tests, T-uptake and ferritin. Clin Biochem 1995;28:181–5.
	3	DPC IMMULITE® 1000 reference ranges used at Children's National Medical Center, Washington, DC.
	4	Dugaw KA, Jack RM, Rutledge J. Pediatric reference ranges for FSH and LH on the Vitros ECi analyzer. Clin Chem 2001;47:A108. (Abstract)
Method(s)	1	IMx (Abbott Laboratories, Abbott, Park, IL).
	2	Immuno I (Bayer Corp., Tarrytown, NY).
	3	DPC IMMULITE® 1000 analyzer (Diagnostic Products Corporation, Los Angeles, CA).
	4	Chemiluminescent immunoassay, Vitros ECi (Ortho-Clinical Diagnostics, Raritan, NJ).
Comment(s)	1,2	Study used hospitalized patients and a computerized approach to removing outliers. Values are 2.5–97.5th percentiles.
	3	*Numbers not provided.
		Pediatric reference ranges were determined by using regression equations to transform the previously established ranges on the Bayer Immuno 1 (Bayer Corp., Tarrytown, NY) based on the Hoffmann technique to their corresponding values on the IMMULITE® 1000.
	4	**Ages ranged from 1 h to 18 y with a total of 92 specimens.

MAGNESIUM

		Male			Female		
Test	**Age**	**n**	**mg/dL**	**mmol/L**	**n**	**mg/dL**	**mmol/L**
1	1–30 d	68	1.7–2.4	0.70–0.99	18	1.7–2.5	0.70–1.03
	31–365 d	62	1.6–2.5	0.66–1.03	37	1.9–2.4	0.78–0.99
	1–3 y	140	1.7–2.4	0.70–0.99	103	1.7–2.4	0.70–0.99
	4–6 y	120	1.7–2.4	0.70–0.99	63	1.7–2.2	0.70–0.91
	7–9 y	118	1.7–2.3	0.70–0.95	102	1.6–2.3	0.66–0.95
	10–12 y	111	1.6–2.2	0.66–0.91	74	1.6–2.2	0.66–0.91
	13–15 y	84	1.6–2.3	0.66–0.95	113	1.6–2.3	0.66–0.95
	16–18 y	73	1.5–2.2	0.62–0.91	92	1.5–2.2	0.62–0.91
2	Premature 0–6 d	48	1.6–2.7	0.65–1.10	48	1.6–2.7	0.65–1.10
	Newborns 0–6 d	134	1.2–2.6	0.48–1.05	134	1.2–2.6	0.48–1.05
	Premature 7–30 d	24	1.8–2.4	0.75–1.00	24	1.8–2.4	0.75–1.00
	Newborn 7–30 d	89	1.6–2.4	0.65–1.00	89	1.6–2.4	0.65–1.00
	1 mo–1 y	164	1.6–2.6	0.65–1.05	164	1.6–2.6	0.65–1.05
	1–2 y	102	1.6–2.6	0.65–1.05	102	1.6–2.6	0.65–1.05
	2–6 y	138	1.5–2.4	0.60–1.00	138	1.5–2.4	0.60–1.00
	6–10 y	130	1.6–2.3	0.65–0.95	130	1.6–2.3	0.65–0.95
	10–14 y	133	1.6–2.2	0.65–0.90	133	1.6–2.2	0.65–0.90
	14 y +	139	1.5–2.3	0.60–0.95	139	1.5–2.3	0.60–0.95
3	1–30 d	*	1.3–2.7	0.5–1.1	*	1.3–2.7	0.5–1.1
	1 mo–24 mo		1.7–2.7	0.7–1.1		1.7–2.7	0.7–1.1
	2–5 y		1.6–2.5	0.7–1.0		1.6–2.5	0.7–1.0
	6–14 y		1.7–2.4	0.7–1.0		1.7–2.4	0.7–1.0
	15–18 y		1.6–2.4	0.7–1.0		1.6–2.3	0.7–0.9

Specimen Type(s)	1,2	Plasma
	3	Plasma/serum
Reference(s)	1	Hicks JM, Bailey J, Bjorn S, et al. Pediatric reference ranges for plasma magnesium. Clin Chem 1995;41:S93. (Abstract)
	2	Meites S, ed. Pediatric clinical chemistry, 3rd ed. Washington, DC: AACC Press, 1989:191.
	3	Ghoshal AK, Soldin SJ. Evaluation of the Dade Behring Dimension RxL: integrated chemistry system-pediatric reference ranges. Clin Chim Acta 2003;331:135–46.
Method(s)	1	Magnesium reacts with calmagite to form a reddish violet chromophore. Hitachi 747 with Boehringer Mannheim reagents (Boehringer Mannheim Diagnostics, Indianapolis, IN).
	2	Colorimetric using a formazan dye. Vitros 700 (Ortho-Clinical Diagnostics, Raritan, NJ).
	3	Methylthymol blue forms a blue complex with magnesium. Calcium interference is minimized by forming a complex between calcium and Ba-EGTA (chelating agent). Dade Behring Dimension RxL Analyzer (Dade Behring Inc., Newark, DE).
Comment(s)	1	Study used hospitalized patients and a computerized approach adapted from the Hoffmann technique to obtain 2.5–97.5th percentiles.
	2	Results are 5–95th percentiles from patients at Children's Hospital, Columbus, OH. Males and females studied together.
	3	*Numbers not provided.
		Reference ranges were obtained by comparing results from previously published data and using regression equations.

MANGANESE

Male and Female			
Age	n	µg/L	nmol/L
0–<0.5 y	13	0.5–3.8	8.7–68.7
0.5–<1.0 y	18	0.6–3.7	10.7–66.3
1.0–<2.0 y	15	0.0–3.4	0.0–61.1
2.0–<4.0 y	23	0.4–2.3	8.1–42.1
4.0–<6.0 y	19	0.2–2.7	3.3–49.3
6.0–<10.0 y	25	0.3–2.3	5.2–40.8
10.0–<14.0 y	21	0.4–2.0	7.6–36.4
14.0–<18.0 y	17	0.4–1.5	7.8–27.8

Specimen Type(s)	Plasma/serum
Reference(s)	Rükgauer M, Klein J, Kruse-James JD. Reference values for the trace elements copper, manganese, selenium, and zinc in the serum/plasma of children, adolescents, and adults. J Trace Elements Med Biol 1997;11:92–8.
Method(s)	Atomic absorption spectrophotometry with Zeeman background compensation. Perkin Elased ETAAS, Zeeman 3030 (Uberlingen, Germany).
Comment(s)	Study population was drawn from patients visiting the outpatient department or surgical or orthopedic ward for preoperative workup. Results are mean ±2 SDs. (2.5th–97.5th percentiles)

METANEPHRINE (URINE)

Male and Female				
Age	n	µg/24h	µmol/24h	µg/g creatinine
3–8 y	76	5–113	0.025–0.514	47–240
9–12 y	44	21–154	0.107–0.782	40–220
13–17 y	43	32–167	0.162–0.848	33–145
Adults	51	45–290	0.228–1.472	31–140

Specimen Type(s)	Urine
Reference(s)	Pediatric endocrine testing. Nichols Institute, 1993:30.
Method(s)	HPLC electrochemical detection.
Comment(s)	Results are 2.5–97.5th percentiles.

METHYLMALONIC ACID

Age	n	μmol/L
		Male and Female
4–8 y	19	0.06–0.24
8–14 y	20	0.03–0.26

Specimen Type(s)	Serum
Reference(s)	Straczek J, Felden F, Dousset B. Quantification of methylmalonic acid in serum measured by capillary gas chromatography-mass spectrometry as tert-butyldimethylsilyl derivates. J Chromat 1993:620;1–7.
Method(s)	Gas chromatography-mass spectrometry.
Comment(s)	Results are 2.5–97.5th percentiles.

β_2-MICROGLOBULIN

	Male		Female	
Age	**n**	**μg/L**	**n**	**μg/L**
1–30 d	68	1603–4790	50	1722–4547
31–182 d	73	1423–3324	68	1024–3774
183–365 d	39	897–3095	27	999–2282
1–3 y	158	827–2228	129	742–2396
4–6 y	142	567–2260	123	546–2170
7–9 y	97	772–1712	79	736–1766
10–12 y	92	699–1836	68	704–1951
13–15 y	103	681–1954	82	787–1916
16–18 y	54	724–1874	77	555–1852

Specimen Type(s)	Plasma/serum
Reference(s)	Soldin SJ, Hicks JM, Bailey J, et al. Pediatric reference ranges for β2-microglobulin and ceruloplasmin. Clin Chem 1997;43:S199. (Abstract)
Method(s)	RIA. Pharmacia β2-Micro RIA (Pharmacia, Uppsala, Sweden).
Comment(s)	Study used hospitalized patients and a computerized approach adapted from the Hoffmann technique. Values are the 2.5–97.5th percentiles.

MUCOPOLYSACCHARIDES (URINE)

Male and Female		
Age	**n**	**mg/mmol creatinine**
0–<6 mo	102	15.2–52.0
6 mo–<12 mo	34	15.1–31.5
1–<2 y	43	9.1–29.9
2–<4 y	55	7.7–21.3
4–<6 y	45	7.6–14.4
6–<8 y	30	5.7–12.9
8–<10 y	16	5.2–11.6
10–<15 y	27	2.4–10.6
15–<20 y	10	1.5–6.7
>20 y	31	1.5–5.1

Specimen Type(s)	Urine
Reference(s)	de Jong JGN, Wevers RA, Liebrand-van Sambeek R. Measuring urinary glycosaminoglycans in the presence of protein: An improved screening procedure for mucopolysaccharides based on dimethyl-methylene blue. Clin Chem 1992; 36:803–7.
Method(s)	Dimethylene blue-Tris assay.
Comment(s)	Results are the 2.5–97.5th percentiles

NOREPINEPHRINE/NORADRENALINE (PLASMA)

Male and Female				
Test	**Age**	**n**	**nmol/L**	**pg/mL**
1	2–10 d	21	1.0–7.0	169–1184
	10 d–3 mo	10	2.2–12.3	372–2081
	3 mo–1 y	14	1.6–6.6	271–1117
	1–2 y	13	0.4–10.7	68–1810
	2–3 y	8	1.1–8.7	186–1472
	3–15 y	20	0.5–7.4	85–1252
2	30 min after birth	16	2.5–9.1	422–1538
	2 h after birth	16	3.1–6.5	532–1108
	3 h after birth	16	0.7–8.5	112–1436
	12 h after birth	16	1.2–1.9	195–327
	24 h after birth	16	1.5–2.8	250–470
	48 h after birth	16	1.3–2.0	221–345

Specimen Type(s)	1,2	Plasma
Reference(s)	1	Candito M, Albertini M, Politano S, et al. Plasma catecholamine levels in children. J Chrom Biomed Appl 1993;617:304–7.
	2	Eliot RJ, Lam R, Leake RD, et al. Plasma catecholamine concentrations in infants at birth and during the first 48 hours of life. J Pediatrics 1980;96:311–5.
Method(s)	1	High performance liquid chromatography.
	2	Radioenzymatic method.
Comment(s)	1	Study population consisted of 86 healthy children (62 males, 24 females) aged 2 days to 15 years.
	2	Study performed on 16 term vaginally delivered infants. Results are mean ±2 SDs.

NOREPINEPHRINE/NORADRENALINE (URINE)

Male and Female				
Test	**Age**	**n**	**µg/g creatinine**	**mmol/mol creatinine**
1	0–24 mo	24	<420	<0.280
	2–4 y	37	<120	<0.080
	5–9 y	40	<89	<0.059
	10–19 y	41	<82	<0.055
2	<1 y	18	25–310	0.017–0.207
	1–4 y	24	25–290	0.017–0.194
	4–10 y	23	27–108	0.018–0.072
	10–18 y	20	4–105	0.003–0.070

Specimen Type(s)	1,2	Urine
Reference(s)	1	Soldin SJ, Lam G, Pollard A, et al. High performance liquid chromatographic analysis of urinary catecholamines employing amperometric detection: Reference values and use in laboratory diagnosis of neural crest tumors. Clin Biochem 1980;13:285–91.
	2	Rosano TG. Liquid chromatographic evaluation of age related changes in the urinary excretion of free catecholamines in pediatric patients. Clin Chem 1984;30:301–3.
Method(s)	1,2	HPLC with electrochemical detection.
Comment(s)	1	Study involved healthy children. Results quoted are 95th percentile.
	2	Urine was obtained from 85 pediatric patients (diagnosis of neoplasia excluded). Results are 0–100th percentiles.

NORMETANEPHRINE (URINE)

Male and Female				
Age	**n**	**µg/24h**	**µmol/24h**	**µg/g creatinine**
3–8 y	76	13–252	0.071–1.375	62–705
9–12 y	44	32–346	0.175–1.888	81–583
13–17 y	43	63–402	0.344–2.194	95–375
Adults	51	82–500	0.448–2.729	47–310

Specimen Type(s)	Urine
Reference(s)	Pediatric endocrine testing. Nichols Institute, 1993:30.
Method(s)	HPLC electrochemical detection.
Comment(s)	Results are 2.5–97.5th percentiles.

OSMOLALITY

Male and Female			
Test	**Age**	**n**	**mOsm/Kg**
1	Birth	44	275–300
	7 d	17	276–305
	28 d	17	274–305
2	1 mo–18 y	*	280–300

Specimen Type(s)	1,2	Serum
Reference(s)	1	Davies DP. Plasma osmolality and protein intake in preterm infants. Arch Dis Child 1973;48:575–9. Meites S, ed. Pediatric clinical chemistry, 3rd ed. Washington, DC: AACC Press, 1989:202–3.
	2	Children's National Medical Center. Unpublished data.
Method(s)	1,2	Freezing-point depression.
Comment(s)	1	The report is based on Davies' study of 53 preterm infants with birthweights <2.5 kg.
	2	*Numbers not provided.

OXYGEN, PARTIAL PRESSURE (pO_2)

Male and Female			
Age	**n**	**mmHg**	**kPa**
Birth	*	8–24	1.1–3.2
5–10 min		33–75	4.4–10.0
30 min		31–85	4.1–11.3
>1 h		55–80	7.3–10.6
1 d		54–95	7.2–12.6
>1 d		83–108	11.0–14.4

Specimen Type(s)	Arterial whole blood
Reference(s)	Behrman RE, ed. Nelson textbook of pediatrics, 14th ed. Philadelphia, PA: WB Saunders Company, 1992:1818.
Method(s)	Not given.
Comment(s)	*Numbers not provided.

OXYGEN SATURATION

Male and Female			
Age	n	%	Saturated Fraction
Newborn	*	85–90	0.85–0.90
Thereafter		95–99	0.95–0.99

Specimen Type(s)	Arterial whole blood
Reference(s)	Behrman RE, ed. Nelson textbook of pediatrics, 14th ed. Philadelphia, PA: WB Saunders Company, 1992:1818.
Method(s)	Not given.
Comment(s)	*Numbers not provided.

PARATHYROID HORMONE (PTH)

Male and Female		
Age	n	pg/mL
Mid-Molecule (C-Terminal)		
1–16 y	39	51–217
N-Terminal Specific		
2–13 y	18	14–21
Intact (IRMA)		
Cord blood	20	≤3.0
2–20 y	150	9–52

Specimen Type	Serum
Reference	Pediatric endocrine testing. Nichols Institute, 1993:31–2.
Method(s)	IRMA (immunoradiometric assay), N-terminal specific (radioimmunoassay), C-terminal (radioimmunoassay).
Comments	Results are 2.5–97.5th percentiles.

pH

Test	Age	n	Male and Female
1	0–1 mo	200	7.18–7.51
	1–6 mo	100	7.18–7.50
	6–12 mo	85	7.27–7.49
2	Cord blood[a]	169	7.26–7.50
	2–5 d[b]	28	7.30–7.49

Specimen Type(s)	1,2	Whole blood
Reference(s)	1	Snell J, Greeley C, Colaco A, et al. Pediatric reference ranges for arterial pH, whole blood electrolytes and glucose. Clin Chem 1993;39:1173. (Abstract)
	2	[a]Dickman KA. In: Meites S, ed. Pediatric clinical chemistry, 3rd ed. Washington, DC: AACC Press, 1989:24–5.
		[b]Lockitch G, Halstead AC. In: Meites S, ed. Pediatric clinical chemistry, 3rd ed. Washington, DC: AACC Press, 1989:25.
Method(s)	1	Ion specific electrode. (Ciba Corning 288 blood gas system. Ciba Corning Diagnostics, East Walpole, MA).
	2	[a]Ion specific electrode. (IL-813, Instrumentation Laboratories, Lexington, MA).
		[b]Ion specific electrode. (Nova Biomedical Stat Profile, Nova Biomedical, Waltham, MA).
Comment(s)	1	Study used hospitalized patients and a computerized approach to removing outliers. Values are 2.5–97.5th percentiles.
	2	[a]In–house study of newborns with Apgar scores of 7 or more.
		[b]Healthy term infants. Mean birth weight 3465 g. Results are 2.5–97.5th percentiles.

PHOSPHORUS

		Male			Female		
Test	Age	n	mg/dL	mmol/L	n	mg/dL	mmol/L
1	1–30 d	62	3.9–6.9	1.25–2.25	66	4.3–7.7	1.40–2.50
	31–365 d	83	3.5–6.6	1.15–2.15	66	3.7–6.5	1.20–2.10
	1–3 y	126	3.1–6.0	1.00–1.95	119	3.4–6.0	1.10–1.95
	4–6 y	112	3.3–5.6	1.05–1.80	107	3.2–5.5	1.05–1.80
	7–9 y	117	3.0–5.4	0.95–1.75	107	3.1–5.5	1.00–1.80
	10–12 y	135	3.2–5.7	1.05–1.85	115	3.3–5.3	1.05–1.70
	13–15 y	109	2.9–5.1	0.95–1.65	110	2.8–4.8	0.90–1.55
	16–18 y	95	2.7–4.9	0.85–1.60	122	2.5–4.8	0.80–1.55
2	0–5 d (<2.5 kg)	50	4.6–8.0	1.50–2.60	50	4.6–8.0	1.50–2.60
	1–3 y	50	3.9–6.5	1.25–2.10	50	3.9–6.5	1.25–2.10
	4–6 y	38	4.0–5.4	1.30–1.75	38	4.0–5.4	1.30–1.75
	7–9 y	72	3.7–5.6	1.20–1.80	72	3.7–5.6	1.20–1.80
	10–11 y	62	3.7–5.6	1.20–1.80	62	3.7–5.6	1.20–1.80
	12–13 y	73	3.3–5.4	1.05–1.75	73	3.3–5.4	1.05–1.75
	14–15 y	91	2.9–5.4	0.95–1.75	91	2.9–5.4	0.95–1.75
	16–19 y	107	2.8–4.6	0.90–1.50	107	2.8–4.6	0.90–1.50
3	0–30 d	181	2.7–7.2	0.87–2.33	140	3.0–8.0	0.97–2.58
	31–90 d	84	3.0–6.8	0.97–2.20	87	3.0–7.5	0.97–2.42
	3–12 mo	109	3.0–6.9	0.97–2.23	119	2.5–7.0	0.81–2.26
	13–24 mo	69	2.5–6.4	0.81–2.07	78	3.0–6.5	0.97–2.10
	2–<13 y	148	3.0–6.0	0.97–1.94	254	2.5–6.0	0.81–1.94
	13–<16 y	175	3.0–5.4	0.97–1.74	72	3.0–5.6	0.97–1.81
	16–<18 y	72	3.0–5.2	0.97–1.68	196	3.0–4.8	0.97–1.55
4	0–30 d	*	2.8–7.0	0.90–2.26	*	3.1–7.7	1.00–2.49
	31–90 d		3.1–6.6	1.00–2.13		3.1–7.2	1.00–2.32
	3–12 mo		3.1–6.6	1.00–2.13		3.1–6.8	1.00–2.20
	13–24 mo		3.1–6.2	1.00–2.00		3.1–6.3	1.00–2.03
	2–<13 y		3.1–5.9	1.00–1.90		3.1–5.9	1.00–1.90
	13–<16 y		3.1–5.3	1.00–1.71		3.1–5.5	1.00–1.78
	16–18 y		3.1–5.1	1.00–1.65		3.1–4.8	1.00–1.55

Specimen(s)	1–4	Plasma/serum
Reference(s)	1	Soldin SJ, Hicks JM, Bailey J, et al. Pediatric reference ranges for phosphate on the Hitachi 747 analyzer. Clin Chem 1997;43:S198. (Abstract)
	2	Lockitch G, Halstead AC, Albersheim S, et al. Age- and sex-specific pediatric reference intervals for biochemistry analytes as measured with the Ektachem-700 analyzer. Clin Chem 1988;34:1622–5.
	3	Soldin SJ, Hunt C, Hicks JM. Pediatric reference ranges for Phosphorus on the Vitros 500 Analyzer. Clin Chem 1999;45:A22. (Abstract)
	4	Ghoshal AK, Soldin SJ. Evaluation of the Dade Behring Dimension RxL: integrated chemistry system-pediatric reference ranges. Clin Chim Acta 2003;331:135–46.
Method(s)	1	Hitachi 747 using ammonium molybdate method (Boehringer Mannheim, Diagnostics, Indianapolis, IN).
	2,3	Vitros 700 (2) and 500 (3) using ammonium molybdate method (Ortho-Clinical Diagnostics, Raritan, NJ).
	4	The phosphorus method uses a mixture of p-methylaminophenol sulfate and bisulfite to reduce the phosphomolybdate. Dade Behring Dimension RxL Analyzer (Dade Behring Inc., Newark, DE).
Comment(s)	1,3	Study used hospitalized patients and a computerized approach to removing outliers. Values are 2.5–97.5th percentiles.
	2	Study used normal healthy children. Values are 2.5–97.5th percentiles.
	4	*Numbers not provided.
		Reference ranges were obtained by comparing results from previously published data and using regression equations.

POTASSIUM

Male and Female			
Test	Age	n	mmol/L
1	0–1wk	100	3.2–5.5
	1 wk–1 mo	100	3.4–6.0
	1–6 mo	100	3.5–5.6
	6 mo–1 y	100	3.5–6.1
	Over 1 y	105	3.3–4.6
2	1–15 y	*	3.7–5.0
	16 y–Adult		3.7–4.8
3	0–1 mo	207	2.5–5.4
	1–6 mo	96	2.7–5.2
4	0–1 wk	**	3.2–5.7
	1 wk–1 mo		3.4–6.2
	1–6 mo		3.5–5.8
	6 mo–1 y		3.5–6.3
	>1 y		3.3–4.7

Specimen Type(s)	1,2	Plasma
	3	Whole blood
	4	Plasma/serum
Reference(s)	1	Greeley C, Snell J, Colaco A, et al. Pediatric reference ranges for electrolytes and creatinine. Clin Chem 1993;39:1172. (Abstract)
	2	Burritt MF, Slockbower JM, Forsman BS, et al. Pediatric reference intervals for 19 biologic variables in healthy children. Mayo Clinic Proceedings 1990;65:329–36.
	3	Snell J, Greeley C, Colaco A, et al. Pediatric reference ranges for arterial pH, whole blood electrolytes, and glucose. Clin Chem 1993;39:1173. (Abstract)
	4	Ghoshal AK, Soldin SJ. Evaluation of the Dade Behring Dimension RxL: integrated chemistry system-pediatric reference ranges. Clin Chim Acta 2003;331:135–46.
Method(s)	1	Vitros 500 (Ortho-Clinical Diagnostics, Raritan, NJ).
	2	Flame Photometry–American Monitor (American Diagnostics, Inc., Indianapolis, IN).
	3	288 Blood Gas System (Ciba Corning Diagnostics, East Walpole, MA).
	4	The sodium, potassium, and chloride (Na/K/Cl) methods use indirect sample sensing with the QuikLYTE® Integrated Multisensor Technology (IMT) to develop an electrical potential proportional to the activity of each specific ion in the sample. The total carbon dioxide (TCO_2) method uses a Severinghaus electrode designed to measure the liberated CO_2 from an acidified sample. Dade Behring Dimension RxL Analyzer (Dade Behring Inc., Newark, DE).
Comment(s)	1	Study used hospitalized patients and a computerized approach to removing outliers. Above values are 2.5–97.5th percentiles.
	2	From normal healthy children. *See reference for numbers.
	3	Study used hospitalized patients and a computerized approach to removing outliers. Above values are 2.5–97.5th percentiles.
	4	**Numbers not provided. Reference ranges were obtained by comparing results from previously published data and using regression equations. Above values are 2.5–97.5th percentiles.

PREALBUMIN (TRANSTHYRETIN)

		Male and Female		
Test	Age	n	mg/L	mg/dL
1	0–1 mo	63	70–390	7.0–39.0
	1–6 mo	55	83–340	8.3–34.0
	6 mo–4 y	159	20–360	2.0–36.0
	4–6 y	59	120–300	12.0–30.0
	6–19 y	189	120–420	12.0–42.0
2	0–5 d	69	60–210	6.0–21.0
	1–5 y	68	140–300	14.0–30.0
	6–9 y	68	150–330	15.0–33.0
	10–13 y	61	200–360	20.0–36.0
	14–19 y	70	220–450	22.0–45.0
3	0–4 d	118	73–144	7.3–14.4
	1 mo–4 y	116	67–171	6.7–17.1
	5–11 y	149	91–220	9.1–22.0
	12–20 y	207	124–302	12.4–30.2
4	0–5 d	*	86–232	8.6–23.2
	6 d–24 mo		96–320	9.6–32.0
	2–5 y		164–320	16.4–32.0
	6–9 y		173–349	17.3–34.9
	10–13 y		222–378	22.2–37.8
	14–19 y		242–466	24.2–46.6

		Male			Female		
Test	Age	n	mg/L	mg/dL	n	mg/L	mg/dL
5	1–40 d	156	32–159.0	3.2–15.9	104	42–144	4.2–14.4
	41–90 d	134	27–176.0	2.7–17.6	90	25–219	2.5–21.9
	3–9 mo	114	73–279.0	7.3–27.9	94	53–250.0	5.3–25.0
	10–24 mo	106	67–285.0	6.7–28.5	80	73–337	7.3–33.7
	2–10 y	325	69–312.0	6.9–31.2	138	80–352	8.0–35.2
	11–15 y	153	63–335.0	6.3–33.5	169	86–407	8.6–40.7
	16–18 y	67	80–416.0	8.0–41.6	89	137–441	13.7–44.1

Specimen Type(s)	1–3,5	Serum
	4	Plasma/serum
Reference(s)	1	Davis ML, Austin C, Messmer BL, et al. IFCC-standardized pediatric reference intervals for 10 serum proteins using the Beckman Array 360 system. Clin Biochem 1996;29:489–92.
	2	Lockitch G, Halstead AZ, Quigley G, et al. Age- and sex-specific pediatric reference intervals: study design and methods illustrated by measurement of serum proteins with the Behring LN nephelometer. Clin Chem 1988;34:1618–21.
	3	Hamlin CR, Pankowsky DA. Turbidimetric determination of transthyretin (prealbumin) with a centrifugal analyzer. Clin Chem 1987;33:144–6.
	4	Ghoshal AK, Soldin SJ. Evaluation of the Dade Behring Dimension RxL: integrated chemistry system-pediatric reference ranges. Clin Chim Acta 2003;331:135–46.
	5	Soldin S, Zhang M, Schaub JR, Ghoshal A. Pediatric reference ranges for prealbumin and retinol binding protein. Clin Biochem 2003 (in press).
Method(s)	1	Rate nephelometry. Beckman Array 360 (Beckman Instruments, Brea, CA).
	2,5	Nephelometry using the Behring LN nephelometer (Behring Diagnostics, Westwood, MA).
	3	Immunoturbidimetric procedure using a Cobas-Bio centrifugal analyzer (Roche Analytical Instruments, Inc., Nutley, NJ) and rabbit antiserum (Behring Diagnostics, La Jolla, CA).
	4	The prealbumin method is based on a particle-enhanced turbidimetric immunoassay (PETIA) technique. Dade Behring Dimension RxL Analyzer (Dade Behring Inc., Newark, DE).
Comment(s)	1	Samples were obtained from children attending outpatient clinics. Results are 2.5–97.5th percentiles.
	2	No significant differences were found for males and females.
		These ranges were therefore derived from combined data. The study used healthy children. Results are 2.5–97.5th percentiles.
	3	Healthy children. Values are 2.5–97.5th percentiles.
	4	*Numbers not provided
		Reference ranges were obtained by comparing results from previously published data and using regression equations.
	5	Study used hospitalized patients in a computerized approach to removing outliers.

PREGNENOLONE

Male and Female			
Age	**n**	**ng/dL**	**nmol/L**
Premature infants	*		
26–28 wks, day 4		260–2104	8.2–66.4
Newborn			
1–7 d		150–2000**	4.7–63.1**
Infants and prepubertal children		20–140	0.6–4.4
Pubertal age groups (11–16 y)		10–150	0.3–4.7

Specimen Type(s)	Serum
Reference(s)	Endocrinology expected values. Esoterix Endocrinology, Calabasas Hills, CA. ©2002 Esoterix, Inc. http://www.esoterix.com
Method(s)	Not provided.
Comment(s)	*Numbers not provided. **Levels decrease after birth, and are within the prepubertal range by three months.

17-OH-PREGNENOLONE

Male and Female			
Age	**n**	**ng/dL**	**nmol/L**
Premature infants	*		
26–28 wks, day 4		375–3559	11.3–106.9
31–35 wks, day 4		64–2380	1.9–71.5
Full-term infants			
3 d		10–829	0.3–24.9
1–6 mo		36–763	1.1–22.9
6–12 mo		42–540	1.3–16.2
Prepubertal children			
1–2 y		14–207	0.4–6.2
2–6 y		10–103	0.3–3.1
6–10 y		10–186	0.3–5.6
Pubertal age groups (11–16 y)		44–235	1.3–7.1

Specimen Type(s)	Serum
Reference(s)	Endocrinology expected values. Esoterix Endocrinology, Calabasas Hills, CA. ©2002 Esoterix, Inc. http://www.esoterix.com
Method(s)	Not provided.
Comment(s)	*Numbers not provided.

PROGESTERONE

		Male			Female		
Test	**Age**	**n**	**ng/dL**	**nmol/L**	**n**	**ng/dL**	**nmol/L**
1	6–9 y	23*	≤20	≤0.64	23*	≤20	≤0.64
	10–11 y	25*	≤20	≤0.64	25*	≤20	≤0.64
	12–17 y	24	≤20	≤0.64			

Male and Female				
Test	**Age**	**n**	**ng/dL**	**nmol/L**
2	Premature infants	**		
	26–28 wks		18–640	0.57–20.35
	31–35 wks		84–1360	2.67–43.25
	Full-term infants			
	1–7 d		[a]	[a]
	Prepubertal children			
	1–10 y		7–52	0.22–1.65

Male			Female[b]		
Age	**ng/dL**	**nmol/L**	**Age**	**ng/dL**	**nmol/L**
Puberty[c]			Puberty[c]		
Tanner Stage[d]			Tanner Stage[d]		
1 (<9.8 y)	<10–33	<0.32–1.06	1 (<9.2 y)	<10–33	<0.32–1.06
2 (9.8–14.5 y)	<10–33	<0.32–1.06	2 (9.2–13.7 y)	<10–55	<0.32–1.76
3 (10.7–15.4 y)	<10–48	<0.32–1.54	3 (10.0–14.4 y)	10–450	0.32–14.4
4 (11.8–16.2 y)	10–108	0.32–3.46	4 (10.7–15.6 y)	10–1300	0.32–41.6
5 (12.8–17.3 y)	21–82	0.67–2.62	5 (11.8–18.6 y)	10–950	0.32–30.4
n**			**n****		

Specimen Type(s)	1,2	Serum
Reference(s)	1	Pediatric endocrine testing. Nichols Institute, 1993:34.
	2	Endocrinology expected values. Esoterix Endocrinology, Calabasas Hills, CA. ©2002 Esoterix, Inc. http://www.esoterix.com
		[c]Data adapted from Bidlingmaier F, Wagner-Barnack M, Butenandt O, Knorr D. Plasma estrogens in childhood and puberty under physiologic and pathologic conditions. Pediat Res 1973;7:901.
		[d]Tanner JM, Whitehouse RH. Clinical longitudinal standards for height, weight, height velocity, and the stages of puberty. Arch Dis Childhood 1976;51:170.
Method(s)	1	Extraction, radioimmunoassay.
	2	Not provided.
Comment(s)	1	Results are 2.5–97.5th percentiles.
		*Males and females combined
	2	**Numbers not provided.
		[a]Progesterone levels are markedly elevated in the neonate, but fall rapidly to reach prepubertal levels by seven days, where they remain until puberty.
		[b]Day of cycle not determined for prepubertal females.
		[d]As used herein, Tanner stages in males encompass development of both pubic hair and genitalia. In females, each stage encompasses both pubic hair and breast development. While this expands the chronological age range for each stage of pubertal development, it results in a better correlation with hormonal values.

PROLACTIN

		Male		Female	
Test	**Age**	**n**	**ng/mL (μg/L)**	**n**	**ng/mL (μg/L)**
1	0–1 mo	45	3.7–81.2	43	0.3–95.0
	1–12 mo	99	0.3–28.9	89	0.2–29.9
	1–3 mo	141	2.3–13.2	114	1.0–17.1
	4–6 y	106	0.8–16.9	139	1.6–13.1
	7–9 y	105	1.9–11.6	98	0.3–12.9
	10–12 y	109	0.9–12.9	108	1.9–9.6
	13–15 y	138	1.6–16.6	122	3.0–14.4
	16–18 y	95	2.7–15.2	104	2.1–18.4
2	<2 y	158	2.7–25.0	41	4.2–20.2
	2–5 y	142	1.7–17.0	91	1.0–17.5
	6–10 y	103	0.7–15.8	140	1.2–15.5
	11–20 y	91	2.0–18.2	280	1.5–19.5
3	<2 y	158	2.3–18.3	41	3.3–14.7
	2–5 y	142	1.6–12.5	91	1.0–12.8
	6–10 y	103	0.8–11.6	140	1.2–11.4
	11–20 y	91	1.8–13.3	280	1.4–14.3

Specimen Type(s)	1–3	Plasma/serum
Reference(s)	1	Cook JF, Hicks JM, Godwin ID, et al. Pediatric reference ranges for prolactin. (Abstract.) Clin Chem 1992;38:959.
	2	Soldin SJ, Morales A, Albalos F, et al. Pediatric reference ranges on the Abbott IMx analyzer for FSH, LH, prolactin TSH, T_4, T_3, free T_4, free T_3, T-uptake, IgE and ferritin. Clin Biochem 1995;28:603–6.
	3	Murthy JN, Hicks JM, Soldin SJ. Evaluation of the Technicon Immuno I Random Access Immunoassay Analyzer and calculation of pediatric reference ranges for endocrine tests, T-uptake, and ferritin. Clin Biochem 1995;28:181–5.
Method(s)	1	Hybritech Tandem Prolactin procedure (Hybritech Inc, San Diego, CA).
	2	IMx Analyzer (Abbott Laboratories, Abbott Park, IL).
	3	Immuno I (Bayer Corp., Tarrytown, NY).
Comment(s)	1–3	Studies used hospitalized patients and a computerized approach adapted from the Hoffmann technique. Values are 2.5–97.5th percentiles.

PROTEIN, TOTAL

		Male			Female		
Test	**Age**	**n**	**g/dL**	**g/L**	**n**	**g/dL**	**g/L**
1	1–60 d	62	3.9–7.6	39–76	69	3.4–7.0	34–70
	61–180 d	130	4.1–7.9	41–79	161	3.9–7.6	39–76
	181d–1 y	268	3.9–7.9	39–79	196	4.5–7.8	45–78
2	0–5 d (<2.5 kg)	30	3.8–6.2	38–62	30	3.8–6.2	38–62
	0–5 d (>2.5 kg)	93	5.4–7.0	54–70	93	5.4–7.0	54–70
	1–3 y	50	5.9–7.0	59–70	50	5.9–7.0	59–70
	4–6 y	38	5.9–7.8	59–78	38	5.9–7.8	59–78
	7–9 y	74	6.2–8.1	62–81	74	6.2–8.1	62–81
	10–19 y	332	6.3–8.6	63–86	332	6.3–8.6	63–86
3	1–30 d	68	4.1–6.3	41–63	51	4.2–6.2	42–62
	31–182 d	58	4.7–6.7	47–67	42	4.4–6.6	44–66
	183–365 d	29	5.5–7.0	55–70	186	5.6–7.9	56–79
	1–18 y	652	5.7–8.0	57–80	440	5.7–8.0	57–80
4	1–60 d	*	4.0–7.6	40–76	*	3.6–7.0	36–70
	61–180 d		4.0–7.0	40–70		4.0–7.6	40–76
	181 d–1 y		4.2–7.9	42–79		4.6–7.8	46–78
	1–6 y		6.0–8.0	60–80		6.0–7.8	60–78
	7–9 y		6.3–8.1	63–81		6.3–8.1	63–81
	10–19 y		6.4–8.6	64–86		6.4–8.6	64–86

Specimen Type(s)	1,3	Plasma/serum
	2	Plasma
Reference(s)	1	Soldin SJ, Morse AS. Pediatric reference ranges for albumin and total protein in children <1 year old using the Vitros 500 analyzer. Clin Chem 1998; 44:A15. (Abstract)
	2	Lockitch G, Halstead AC, Albersheim S, et al. Age- and sex-specific pediatric reference intervals for biochemistry analytes as measured with the Ektachem-700 analyzer. Clin Chem 1988;34:1622–5.
	3	Hicks JM, Bjorn S, Beatey J, et al. Pediatric reference ranges for albumin, globulin and total protein on the Hitachi 747. Clin Chem 1995;41:S93. (Abstract)
	4	Ghoshal AK, Soldin SJ. Evaluation of the Dade Behring Dimension RxL: integrated chemistry system-pediatric reference ranges. Clin Chim Acta 2003;331:135–46.
Method(s)	1	Vitros 500 (Ortho-Clinical Diagnostics, Raritan, NJ).
	2	Biuret method. Vitros 700 (Ortho-Clinical Diagnostics, Raritan, NJ).
	3	Boehringer Mannheim total protein reagent (Biuret). Total protein was measured on the Hitachi 747 (Boehringer Mannheim Diagnostics, Indianapolis, IN).
	4	The total protein method is a modification of the biuret reaction. Dade Behring Dimension RxL Analyzer (Dade Behring Inc., Newark, DE).
Comment(s)	1,3	Study used hospitalized patients and a computerized approach to removing outliers. Values are 2.5–97.5th percentiles.
	2	Healthy term and pre-term neonates and normal healthy children.
		Values are 2.5–97.5th percentiles.
	4	*Numbers not provided.
		Reference ranges were obtained by comparing results from previously published data and using regression equations.

PROTEIN CEREBROSPINAL FLUID (CSF PROTEIN)

		Male			Female		
Test	**Age**	**n**	**mg/dL**	**mg/L**	**n**	**mg/dL**	**mg/L**
1	0–14 d	74	<117	<1170	49	<184	<1840
	15–30 d	92	<112	<1120	70	<118	<1180
	31–90 d	223	<88	<880	150	<108	<1080
	3–6 mo	54	<52	<520	53	<46	<460
	7–24 mo	98	<52	<520	77	<54	<540
	2–7 y	101	<54	<540	142	<51	<510
	8–12 y	89	<48	<480	104	<48	<480
	13–18 y	76	<42	<420	85	<48	<480
2	0–14 d	*	15–100	150–1000	*	15–153	150–1530
	15–30 d		15–96	150–960		15–100	150–1000
	31–90 d		15–48	150–480		15–93	150–930
	3–6 mo		15–48	150–480		15–44	150–440
	7–24 mo		15–50	150–500		15–48	150–480
	2–7 y		15–45	15–450		15–45	150–450
	8–12 y		15–40	15–400		15–45	150–450
	13–18 y		15–40	15–400		15–45	150–450

Specimen Type(s)	1,2	Cerebrospinal fluid
Reference(s)	1	Soldin SJ, Baumel CR. Pediatric reference ranges for CSF protein and plasma lactate on the Vitros 500 analyzer. Clin Chem 2001;47:A108. (Abstract)
	2	Ghoshal AK, Soldin SJ. Evaluation of the Dade Behring Dimension RxL: integrated chemistry system-pediatric reference ranges. Clin Chim Acta 2003;331:135–46.
Method(s)	1	Vitros 500 analyzer using Ortho-Clinical Diagnostics reagents (Ortho-Clinical Diagnostics, Raritan, NJ).
	2	Pyrogallol red combines with sodium molybdate to form a red complex with maximum absorbance at 470 nm. The protein in the sample reacts with this complex in acid solution to form a bluish-purple colored complex, which absorbs at 600 nm. Dade Behring Dimension RxL Analyzer (Dade Behring Inc., Newark, DE).
Comment(s)	1	Studies used hospitalized patients and a computerized approach adapted from the Hoffmann technique. Values are 97.5th percentile.
	2	*Numbers not provided. Reference ranges were obtained by comparing results from previously published data and using regression equations.

PYRUVATE

Male and Female				
Test	Age	n	mg/dL	µmol/L
1	All ages	*	0.70–1.32	80–150
2	All ages	*	0.3–0.70	35–80

Specimen Type(s)	1,2	Whole blood, proteins precipitated with perchloric acid.
Reference(s)	1	Reference values and SI unit information. The Hospital for Sick Children, Toronto, 1993:378.
	2	Children's National Medical Center, Washington, DC. Unpublished data.
Method(s)	1,2	Standard enzymatic approach.
Comment(s)	1,2	*Numbers not available.

RENIN (PLASMA)

Male and Female		
Age	n	ng/L
1–6 y	63	6.3–149
7–12 y	68	5.5–110
13–17 y	87	3.3–61

Specimen Type(s)	Plasma
Reference(s)	Coates JE, Chapelski LJ, Yatscoff RW. Pediatric reference intervals for plasma renin. Clin Biochem 1994;27:316–7.
Method(s)	Renin Active Pasteur CT kits (Sanofi Diagnostics Pasteur, Inc., Montreal, Quebec, Canada).
Comment(s)	Children seen as outpatients were tested. Male and female ranges were not significantly different and they were combined for calculations of the above ranges. The results reported are 2.5–97.5th percentiles.

RENIN ACTIVITY

Male and Female		
Age	**n**	**ng/mL/h**
0–1 wk	7	0–40
2–4 wk	7	0–175
3 mo–1 y	*	≤15
1–4 y	*	≤10
4–15 y	*	≤6

Specimen Type(s)	Plasma
Reference(s)	Pediatric endocrine testing. Nichols Institute, 1993:33.
Method(s)	Radioimmunoassay.
Comment(s)	Normal sodium diet. Patients over 1 y of age were supine from 9–11 am before specimens were drawn. *Numbers not provided.

RETINOL BINDING PROTEIN (RBP)

Male and Female				
Test	Age	n	mg/dL	µmol/L
1	0–5 d	64	0.8–4.5	0.38–2.15
	1–5 y	68	1.0–7.6	0.48–3.62
	6–9 y	64	2.0–7.8	0.95–3.72
	10–13 y	59	1.3–9.9	0.62–4.72
	14–19 y	70	3.0–9.2	1.43–4.39
2	Term neonates (>37 wk)	41	1.02–3.14	0.49–1.51
	Preterm neonates (<36.6 wk)	58	0.72–2.88	0.35–1.38

		Male			Female		
Test	Age	n	mg/dL	µmol/L	n	mg/dL	µmol/L
3	0–30 d	114	0.91–5.6	0.43–2.67	92	1.01–5.75	0.48–2.74
	1 mo–18 y	96	1.09–8.7	0.52–4.15	97	0.98–6.76	0.47–3.22

Specimen Type(s)	1	Serum
	2,3	Plasma/serum
Reference(s)	1	Lockitch G, Halstead AZ, Quigley G, et al. Age- and sex-specific pediatric reference intervals: study design and methods illustrated by measurement of serum proteins with the Behring LN nephelometer. Clin Chem 1988;34:1618–21.
	2	Cardona-Pérez A, et al. Cord blood retinol and retinol binding protein in preterm and term neonates. Nutrition Research 1996:16;191–6.
	3	Soldin S, Zhang M, Schaub JR, Ghoshal A. Pediatric reference ranges for prealbumin and retinol binding protein. Clin Biochem 2003 (in press).
Method(s)	1,3	Nephelometry using the Behring LN nephelometer (Behring Diagnostics, Westwood, MA).
	2	Radial Immunodiffusion (Behring Diagnostics, La Jolla, CA).
Comment(s)	1	No significant differences were found for males and females. These ranges were therefore derived from combined data. The study used healthy children. Results are 2.5–97.5th percentiles.
	2	Healthy term and preterm neonates studied. Results are 2.5–97.5th percentiles.
	3	Study used hospitalized patients in a computerized approach to removing outliers.

REVERSE TRIIODOTHYRONINE (rT_3)

Male and Female			
Age	n	ng/dL	nmol/L
1–5 y	*	15–71	0.23–1.10
5–10 y		17–79	0.26–1.20
10–15 y		19–88	0.29–1.36
Adults		30–80	0.46–1.23

Specimen Type(s)	Serum
Reference(s)	Behrman RE, ed. Nelson textbook of pediatrics, 14th ed. Philadelphia, PA: WB Saunders Company, 1992:1821.
Method(s)	Not given.
Comment(s)	*Numbers not provided.

SELENIUM

Male and Female				
Test	Age	n	µg/dL	µmol/L
1	0–5 d*	20	6–9	0.72–1.20
	1–5 y*	30	10–14	1.22–1.82
	6–9 y*	30	10–16	1.29–2.05
	10–14 y*	30	10–19	1.31–2.35
	15–19 y*	27	10–19	1.31–2.35
2	3–5 y	38	2.0–12.4	0.25–1.57
	6–8 y	43	1.6–12.0	0.20–1.52
	9–11 y	52	2.4–13.5	0.30–1.70
	12–13 y	24	4.0–10.7	0.51–1.35
	14–16 y	29	4.0–10.7	0.51–1.35
3	0–<0.5 y	13	0.8–4.4	0.10–0.58
	0.5–<1.0 y	18	0–6.9	0.00–0.87
	1.0–<2.0 y	15	1.5–7.3	0.20–0.96
	2.0–<4.0 y	23	0.9–10.4	0.12–1.36
	4.0–<6.0 y	19	0.6–12.5	0.08–1.64
	6.0–<10.0 y	25	2.4–11.8	0.31–1.55
	10.0–<14.0 y	21	2.4–11.4	0.30–1.50
	14.0–<18.0 y	17	2.1–11.8	0.27–1.55

Specimen Type(s)	1,2	Serum
	3	Plasma/serum
Reference(s)	1	Lockitch G, Halstead A, Wadsworth L, et al. Age- and sex-specific pediatric reference intervals for zinc, copper, selenium, iron, vitamins A and E, and related proteins. Clin Chem 1988;34:1625–8.
	2	Malvy DJ–M, Arnaud J, Burtschy B, et al. Reference values for serum, zinc and selenium of French healthy children. Eur J Epidemiol 1993;9:155–61.
	3	Rükgauer M, Klein J, Kruse-Jarres JD. Reference values for the trace elements copper, manganese, selenium, and sinc in the serum/plasma of children, adolescents, and adults. J Trace Elements Med Biol 1997;11:92–8.
Method(s)	1	Varian GTA-95. Atomic absorption (Varian Canada, Inc., Georgetown, Ontario, Canada).
	2	Perkin-Elmer atomic absorption spectrophotometry (Perkin Elmer Corporation, Norwalk, CT).
	3	Atomic absorption spectrophotometry with Zeeman background compensation. Perkin Elmer ETAAS, Zeeman 3030 (Uberlingen, Germany).
Comment(s)	1	The study population was healthy children. Non-parametric methods were used to determine the 0.025 and 0.975 fractiles.
		*No significant differences were found for males and females. These ranges were therefore derived from combined data.
	2	The study population consisted of healthy French children. Values reported are the 2.5–97.5th percentiles.
		Note: Reference ranges differ greatly on a geographic basis. Data must be interpreted against locally derived reference ranges.
	3	Study population was drawn from patients visiting the outpatient department or surgical or orthopedic ward for preoperative work up. Results are mean ±2 SDs (2.5–97.5th percentiles).

SERUM OSTEOCALCIN

	Male		Female	
Age	n	μg/L	n	μg/L
1 y	43	52–63	31	48–79
2 y	24	64–70	31	61–88
3 y	39	54–70	35	44–93
4 y	43	53–90	47	51–98
5 y	55	52–83	49	66–88
6 y	59	51–84	66	65–96
7 y	51	49–104	42	53–97
8 y	66	51–80	78	50–89
9 y	70	62–83	57	61–78
10 y	86	58–88	82	65–96
11 y	78	56–80	69	36–93
12 y	41	68–80	39	74–99
13 y	47	74–98	35	42–75
14 y	55	66–102	51	25–43
15 y	49	61–105	50	35–59
16 y	27	36–50	39	27–47

Specimen Type(s)	Serum
Reference(s)	Cioffi M, Molinari AM, Gazzero P, et al. Serum osteocalcin in 1634 healthy children. Clin Chem 1997;43:543–5.
Method(s)	Sandwich IRMA (Osteo-ELSA; CIS Bio International, Gif sur Yvette, France).
Comment(s)	Study used 1634 healthy children. Results are 25th and 75th percentiles. Osteocalcin is an important marker of bone turnover.

SEX HORMONE BINDING GLOBULIN (SHBG)

		Male			Female		
Test	**Age**	**n**	**nmol/L**	**µg/dL**	**n**	**nmol/L**	**µg/dL**
1	1–30 d	57	10.8–70.8	0.31–2.04	49	11.8–51.4	0.34–1.48
	31–365 d	93	60.2–208.5	1.73–6.01	68	50.4–181.2	1.45–5.22
	1–3 y	109	42.4–155.6	1.22–4.48	98	51.4–157.7	1.48–4.54
	4–6 y	75	39.4–145.6	1.14–4.20	73	47.8–142.1	1.38–4.10
	7–9 y	70	37.7–114.4	1.09–3.30	29	31.0–103.0	0.89–2.97
	10–12 y	113	31.6–92.5	0.91–2.67	60	20.0–99.6	0.58–2.87
	13–15 y	78	13.3–62.6	0.38–1.80	80	16.6–76.5	0.48–2.20
	16–18 y	34	10.6–53.6	0.31–1.54	39	9.3–75.2	0.27–2.17

Specimen Type(s)	Plasma/serum
Reference(s)	Soldin SJ, Hicks JM, Bailey J, et al. Pediatric reference ranges for sex hormone binding globulin. Clin Chem 1997;43:S200. (Abstract)
Method(s)	Time-resolved fluoroimmunoassay. Delfia SHBG kit (Wallac Oy, Turku, Finland).
Comment(s)	Study used hospitalized patients and a computerized adaptation of the Hoffmann technique. Values are 2.5–97.5th percentiles.

SODIUM

Male and Female			
Test	**Age**	**n**	**mmol/L**
1	0–7 d	100	133–146
	7–31 d	100	134–144
	1–6 mo	100	134–142
	6 mo–1 y	100	133–142
	>1 y	105	134–143
2	0–1 mo	59	127–143
	2–6 mo	49	130–147
3	0–7 d	*	131–144
	7–31 d		132–142
	1–6 mo		132–140
	6 mo–1 y		131–140
	>1 y		132–141

Specimen Type(s)	1	Plasma
	2	Heparinized whole blood
	3	Plasma/serum
Reference(s)	1	Greeley C, Snell J, Colaco A, et al. Pediatric reference ranges for electrolytes and creatinine. Clin Chem 1993;39:1172. (Abstract)
	2	Snell J, Greeley C, Colaco A, et al. Pediatric reference ranges for arterial pH, whole blood electrolytes and glucose. Clin Chem 1993;39:1173. (Abstract)
	3	Ghoshal AK, Soldin SJ. Evaluation of the Dade Behring Dimension RxL: integrated chemistry system-pediatric reference ranges. Clin Chim Acta 2003;331:135–46.
Method(s)	1	Vitros 700 analyzer (Ortho-Clinical Diagnostics, Raritan, NJ).
	2	Corning 288 Blood Gas System (Ciba Corning Diagnostics, East Walpole, MA).
	3	The sodium, potassium, and chloride (Na/K/Cl) methods use indirect sample sensing with the QuikLYTE® Integrated Multisensor Technology (IMT) to develop an electrical potential proportional to the activity of each specific ion in the sample. Dade Behring Dimension RxL Analyzer (Dade Behring Inc., Newark, DE).
Comment(s)	1,2	Study used hospitalized patients and a computerized approach adapted from the Hoffmann technique to obtain the 2.5–97.5th percentiles.
	3	*Numbers not provided.
		Reference ranges were obtained by comparing results from previously published data and using regression equations.

SWEAT ELECTROLYTES

Male and Female				
Test	**Age**	**n**	**mmol/L**	**mmol/L**
1			chloride	sodium
	1 wk–adult	*	<40	<40
	Patients with cystic fibrosis		>60	>60
2	0–3.9 mo	170	6–40	4–40
	4–11.9 mo	70	4–29	5–39
	1–4.9 y	134	6–54	6–54

Specimen Type(s)	1,2	Sweat
Reference(s)	1	Children's National Medical Center, Washington, DC. Unpublished data.
	2	Meites S, ed. Pediatric clinical chemistry, 3rd ed. Washington, DC: AACC Press, 1989:246.
Method(s)	1	Pilocarpine iontophoresis/macroduct collection. Measured on Vitros 500 (Ortho-Clinical Diagnostics, Raritan, NJ). C1 measured on Chloridometer (Radiometer, Cleveland, OH).
	2	Pilocarpine iontophoresis. Flame photometry for sodium, chloridometer for chloride.
Comment(s)	1	Results between 40–60 mmol/L require repeat testing and/or use of other tests such as immunoreactive trypsin, bentiromide test of pancreatic function, and DNA probe test for CF gene.
		*Numbers not provided.
	2	Based on in-house studies of non-affected individuals.

TESTOSTERONE

		Male			Female		
Test	**Age**	**n**	**ng/dL**	**nmol/L**	**n**	**ng/dL**	**nmol/L**
1	1–5 mo	10	1–177	0.03–6.14	5	1–5	0.03–0.17
	6–11 mo	8	2–7	0.07–0.24	6	2–5	0.07–0.17
	1–5 y	16	2–25	0.07–0.87	17	2–10	0.07–0.35
	6–9 y	30	3–30	0.10–1.04	16	2–20	0.07–0.35
	10–11 y	20	5–50	0.17–1.73	10	5–25	0.17–0.87
	12–14 y	28	10–572	0.35–19.83	17	10–40	0.35–1.39
	15–17 y	18	220–800	7.63–27.74	11	5–40	0.17–1.39
2	Premature infants	*			*		
	26–28 wks, day 4		59–125	2.1–4.4		5–16	0.18–0.6
	31–35 wks, day 4		37–198	1.3–6.9		5–22	0.18–0.8
	Full-term infants						
	Newborns		75–400	2.6–14		20–64	0.7–2.2
	1–7 mo[a,b]		[c]	[c]		[d]	[d]
	Prepubertal children						
	1–10 y		<3–10	<0.1–0.4		<3–10	<0.11–0.4

	Male			Female[e]		
Test	**Age**	**ng/dL**	**nmol/L**	**Age**	**ng/dL**	**nmol/L**
2 *cont.*	Puberty[f]			Puberty[f]		
	Tanner Stage[g]			Tanner Stage[g]		
	1 (<9.8 y)	<3–10	<0.1–0.4	1 (<9.2 y)	<3–10	<0.1–0.4
	2 (9.8–14.5 y)	18–150	0.6–5.3	2 (9.2–13.7 y)	7–28	0.3–1.0
	3 (10.7–15.4 y)	100–320	3.5–11.2	3 (10.0–14.4 y)	15–35	0.5–1.2
	4 (11.8–16.2 y)	200–620	7.0–21.7	4 (10.7–15.6 y)	13–32	0.5–1.1
	5 (12.8–17.3 y)	350–970	12.3–34.0	5 (11.8–18.6 y)	20–38	0.7–1.3
	n*			**n***		

Specimen Type(s)	1,2	Serum
Reference(s)	1	Pediatric endocrine testing. Nichols Institute, 1993:38.
	2	Endocrinology expected values. Esoterix Endocrinology, Calabasas Hills, CA. ©2002 Esoterix, Inc. http://www.esoterix.com
		[a]Data adapted from Pang S, Levine LS, Chow D, Sagiani F, Saenger P, New MI. Dihydrotestosterone and its relationship to testosterone in infancy and childhood. J Clin Endocrinol Metab 1979;48:821.
		[b]Data adapted from Forest MG, Cathiard AM, Bertrand JA. Evidence of testicular activity in early infancy. J Clin Endocrinol Metab 1973;37:148.
		[f]Data adapted from Bidlingmaier F, Wagner-Barnack M, Butenandt O, Knorr D. Plasma estrogrens in childhood and puberty under physiologic and pathologic conditions. Pediat Res 1973;7:901.
		[g]Tanner JM, Whitehouse RH. Clinical longitudinal standards for height, weight, height velocity, and the stages of puberty. Arch Dis Childhood 1976;51:170.
Method(s)	1	Extraction, chromatography, radioimmunoassay.
	2	Not provided.
Comment(s)	1	Results are 2.5–97.5th percentiles.
	2	*Numbers not provided.
		[c]Levels decrease rapidly the first week to 20–50 ng/dL, then increase to 60–400 ng/dL (mean = 190) between 20–60 days. Levels then decline to prepubertal range by seven months.
		[d]Levels decrease during the first month to <10 ng/dL and remain there until puberty.
		[e]Day of cycle not determined for pubertal females.
		[g]As used herein, Tanner stages in males encompass development of both pubic hair and genitalia. In females, each stage encompasses both pubic hair and breast development. While this expands the chronological age range for each stage of pubertal development, it results in a better correlation with hormonal values.

THYROID STIMULATING HORMONE (TSH)

		Male		Female	
Test	Age	n	µU/mL mU/L	n	µU/mL mU/L
1	0–1 mo	84	<6.5	62	<6.0
	1–12 mo	114	<4.1	103	<4.0
	1–3 y	128	<3.0	126	<3.3
	4–6 y	109	<3.0	82	<2.8
	7–12 y	112	<3.1	107	<2.9
	13–18 y	106	<3.1	106	<3.0
2	1–30 d	63	0.52–16.00	89	0.72–13.10
	1 mo–5 y	95	0.55–7.10	95	0.46–8.10
	6–18 y	95	0.37–6.00	96	0.36–5.80
3	1–30 d	63	0.76–16.11	89	0.96–13.68
	1 mo–5 y	95	0.80–7.52	95	0.70–8.55
	6–18 y	95	0.61–6.39	96	0.60–6.18

			Male and Female
Test	Age	n	mU/L (µU/mL)
4	0–<2 y	36	0.668–4.465
	2–<7 y	149	0.400–3.200
	7–<13 y	128	0.300–2.689
	13–<18 y	123	0.400–1.900

Test	Age	n	mIU/L
5	0–3 d	*	1.0–20.0
	3–30 d		0.5–6.5
	1–5 mo		0.5–6.0
	6 mo–18 y		0.5–4.5

			Male	Female
Test	Age	n	mIU/L	mIU/L
6	1–30 d	**	0.64–12.75	0.80–10.83
	1 mo–5 y		0.67–5.97	0.59–6.78
	6–18 y		0.52–5.08	0.51–4.91

Specimen Type(s)	1–4,6	Plasma/serum
	5	Serum
Reference(s)	1	Hicks JM, Godwin ID, Beatey J, et al. Pediatric reference ranges for highly sensitive TSH. Clin Chem 1992;38:960. (Abstract)
	2	Soldin SJ, Morales A, Albalos F, et al. Pediatric reference ranges on the Abbott IMx for FSH, LH, prolactin, TSH, T_4, T_3, free T_4, free T_3, T-uptake, IgE and ferritin. Clin Biochem 1995;28:603–6.
	3	Murthy JN, Hicks JM, Soldin SJ. Evaluation of the Technicon Immuno I Random Access Immunoassay Analyzer and calculation of pediatric reference ranges for endocrine tests, T-uptake, and ferritin. Clin Biochem 1995;28:181–5.
	4	Soldin SJ, Hicks JM, Bailey J, et al. Pediatric reference ranges for 3rd generation TSH on the ACS 180. Clin Chem 1998; 44:A13. (Abstract)
	5	Dugaw KA, Jack RM, Rutledge J. Pediatric reference ranges for TSH, free T4, total T4, total T3 and T3 uptake on the Vitros ECi analyzer. Chem Chem 2001;47:A108. (Abstract)
	6	Ghoshal AK, Soldin SJ. Evaluation of the Dade Behring Dimension RxL: integrated chemistry system-pediatric reference ranges. Clin Chim Acta 2003;331:135–46.
Method(s)	1	Delphia Immunofluorescent TSH Kit (Pharmacia ENI Diagnostics Inc., Columbia, MD).
	2	Abbott IMx analyzer (Abbott Laboratories, Abbott Park, IL).
	3	Immuno I (Bayer Corp., Tarrytown, NY).
	4	ACS 180 using Chiron Diagnostics TSH-3kit (Chiron Diagnostics Corp., East Walpole, MA).
	5	Chemiluminescent immunoassay, Vitros ECi (Ortho-Clinical Diagnostics, Raritan, NJ).
	6	The TSH method is a one-step enzyme immunoassay based on the "sandwich" principle. Dade Behring Dimension RxL Analyzer (Dade Behring Inc., Newark, DE).
Comment(s)	1	Study used hospitalized patients and a computerized approach adapted from the Hoffmann technique. Values are 97.5th percentile. Lower limit of detection 0.1 μU/mL.
	2,3	Study used hospitalized patients and a computerized approach adapted from the Hoffmann technique. Values are 2.5–97.5th percentiles. Lower limit of detection 0.03 μU/mL.
	4	Study used hospitalized patients and a computerized approach adapted from the Hoffmann technique. Values are 2.5–97.5th percentiles. Lower limit of detection 0.003 μU/mL.
	5	*Ages ranged from 1 h to 8 y with a total of 119 specimens.
	6	**Numbers not provided.
		Reference ranges were obtained by comparing results from previously published data and using regression equations.

THYROXINE BINDING GLOBULIN (TBG)

		Male		Female	
Test	Age	n	mg/L	n	mg/L
1	1–12 mo	189	16.2–32.9	138	17.7–32.0
	1–3 y	167	16.4–32.0	112	19.3–33.8
	4–6 y	100	16.6–29.8	91	18.3–30.8
	7–12 y	184	16.5–28.8	133	15.0–29.2
	13–18 y	158	13.4–25.6	130	13.7–28.7
2	Cord blood	*	19–39	*	19–39
	1–11 mo		16–36		17–37
	1–9 y		12–28		15–27
	10–19 y		14–26		14–30

Specimen Type(s)	1	Plasma
	2	Plasma/serum
Reference(s)	1	Hicks JM, Godwin ID, Beatey J, et al. Pediatric reference ranges for thyroid binding globulin. Clin Chem 1993:39;1172. (Abstract)
	2	Levy RP, Marshall JS, Velayo NL. Radioimmunoassay of human thyroxine binding globulin (TBG). J Clin Endocrinol Metab 1971:32;372–81.
Method(s)	1	Corning TBG ^{125}I–Radioimmunoassay. (Ciba Corning Diagnostics Corp., East Walpole, MA).
	2	Direct RIA using goat antihuman TBG and ^{125}I–labeled human TBG. A second antibody is used for the bound/free separation phase.
Comment(s)	1	Study used hospitalized patients and a computerized approach to removing outliers. Values are 2.5–97.5th percentiles.
	2	Study used healthy children.
		*For numbers at each age group, see Meites S, ed. Pediatric clinical chemistry, 3rd ed. Washington, DC: AACC Press, 1989:254.

THYROXINE (T_4)

Test	Age	Male n	Male µg/dL	Male nmol/L	Female n	Female µg/dL	Female nmol/L
1	1–30 d	108	3.0–14.4	39–185	116	3.0–13.4	39–172
	1–12 mo	98	5.3–16.3	68–210	95	4.6–13.4	59–172
	1–5 y	151	5.5–11.4	71–147	197	6.3–12.8	81–165
	6–10 y	151	5.4–10.6	69–136	146	5.3–10.8	68–139
	11–15 y	171	4.5–10.3	58–133	160	4.9–10.0	63–129
	16–18 y	141	4.9–8.8	63–113	143	5.1–10.0	66–129
2	1–30 d	100	5.9–21.5	76–276	106	6.3–21.5	81–276
	31–364 d	107	6.4–13.9	82–179	96	4.9–13.7	63–176
	1–3 y	149	7.0–13.1	90–169	132	7.1–14.1	91–180
	4–6 y	139	6.1–12.6	79–162	117	7.2–14.0	93–180
	7–12 y	129	6.7–13.4	86–172	134	6.1–12.1	79–156
	13–15 y	145	4.8–11.5	62–148	108	5.8–11.2	75–144
	16–18 y	49	5.9–11.5	76–148	112	5.2–13.2	67–170
3	1–30 d	108	3.4–12.6	44–163	116	3.4–11.8	44–152
	1–12 mo	98	5.4–14.1	68–182	95	4.7–11.8	60–152
	1–5 y	151	5.3–10.2	69–131	197	6.0–11.3	78–146
	6–10 y	151	5.3–9.5	69–122	146	5.2–9.7	67–125
	11–15 y	171	4.6–9.2	59–119	160	4.9–9.0	63–116
	16–18 y	141	4.9–8.1	63–104	143	5.1–9.0	66–116
4	1–30 d	*	3.4–14.5	44–187	*	3.5–13.5	45–174
	1–12 mo		5.6–16.4	72–211		5.0–13.5	64–174
	1–5 y		5.9–11.6	76–149		6.7–12.9	86–166
	6–10 y		5.7–10.8	73–139		5.6–11.0	72–142
	11–15 y		4.9–10.5	63–135		5.3–10.2	68–131
	16–18 y		5.2–9.1	68–117		5.5–10.2	71–131

Test	Age	Male and Female n	Male and Female µg/dL	Male and Female nmol/L
5	0–3 d	**	8.0–20.0	103–258
	3–30 d		5.0–15.0	64–193
	31–365 d		6.0–14.0	77–180
	1–5 y		4.5–11.0	58–142
	6–18 y		4.5–10.0	58–129

Specimen Type(s)	1–4	Plasma/serum
	5	Serum
Reference(s)	1	Soldin SJ, Morales A, Albalos F, et al. Pediatric reference ranges on the Abbott IMx analyzer for FSH, LH, prolactin, TSH, T_4, T_3, free T_4, free T_3, T-uptake, IgE and ferritin. Clin Biochem 1995;28:603–6.
	2	Soldin SJ, Cook J, Beatey J, et al. Pediatric reference ranges for thyroxine and triiodothyronine uptake. Clin Chem 1992;38:960. (Abstract)
	3	Murthy JN, Hicks JM, Soldin SJ. Evaluation of the Technicon Immuno I Random Access Immunoassay Analyzer and calculation of pediatric reference ranges for endocrine tests, T-uptake, and ferritin. Clin Biochem 1995;28:181–5.
	4	Ghoshal AK, Soldin SJ. Evaluation of the Dade Behring Dimension RxL: integrated chemistry system-pediatric reference ranges. Clin Chim Acta 2003;331:135–46.
	5	Dugaw KA, Jack RM, Rutledge J. Pediatric reference ranges for TSH, free T4, total T4, total T3 and T3 uptake on the Vitros ECi analyzer. Chem Chem 2001;47:A108. (Abstract)
Method(s)	1	Abbott IMx Analyzer (Abbott Diagnostics, Inc., Abbott Park, IL).
	2	T_4: Gamma Coat TM[125] (Baxter-Travenol Diagnostics, Inc., Cambridge, MA).
	3	Immuno I (Bayer Corp., Tarrytown, NY).
	4	The thyroxine method is an adaptation of the EMIT® homogeneous enzyme immunoassay technology. Dade Behring Dimension RxL Analyzer (Dade Behring Inc., Newark, DE).
	5	Chemiluminescent immunoassay, Vitros ECi (Ortho-Clinical Diagnostics, Raritan, NJ).
Comment(s)	1–3	Study used hospitalized patients and a computerized approach adapted from the Hoffmann technique. Values are 2.5–97.5th percentiles.
	4	*Numbers not provided.
		Reference ranges were obtained by comparing results from previously published data and using regression equations.
	5	**Ages ranged from 1 h to 18 y with a total of 119 specimens.

THYROXINE, FREE (FREE T_4)

		Male			Female		
Test	Age	n	ng/dL	pmol/L	n	ng/dL	pmol/L
1	1–3 d	24	0.80–2.78	10–36	38	0.88–1.93	11–25
	4–30 d	73	0.48–2.32	6–30	62	0.61–1.93	8–25
	1–12 mo	52	0.76–2.00	10–26	54	0.88–1.84	11–24
	1–5 y	100	0.90–1.59	12–21	117	1.02–1.72	13–22
	6–10 y	104	0.81–1.68	10–22	101	0.82–1.58	11–20
	11–15 y	101	0.92–1.57	12–20	100	0.79–1.49	10–19
	16–18 y	110	0.92–1.53	12–20	101	0.83–1.44	11–19
2	1–3 d	24	1.16–2.95	15–38	38	1.09–2.09	14–27
	4–30 d	73	0.78–2.25	10–29	62	0.85–2.09	11–27
	1–12 mo	52	1.00–2.17	13–28	54	1.09–2.02	14–26
	1–5 y	100	1.16–1.71	15–22	117	1.24–1.86	16–24
	6–10 y	104	1.09–1.86	14–24	101	1.09–1.71	14–22
	11–15 y	101	1.16–1.71	15–22	100	1.00–1.63	13–21
	16–18 y	110	1.16–1.71	15–22	101	1.09–1.63	14–21

			Male		Female	
Test	Age	n	ng/dL	pmol/L	ng/dL	pmol/L
3	1–3 d	*	0.97–1.87	12.5–24.1	0.93–1.44	12.0–18.5
	4–30 d		0.78–1.52	10.0–19.6	0.81–1.44	10.4–18.5
	1–12 mo		0.89–1.48	11.5–19.1	0.93–1.40	12.0–18.0
	1–5 y		0.97–1.25	12.5–16.1	1.01–1.32	13.0–17.0
	6–10 y		0.93–1.32	12.0–17.0	0.93–1.25	12.0–16.1
	11–15 y		0.97–1.25	12.5–16.1	0.89–1.20	11.5–15.4
	16–18 y		0.97–1.25	12.5–16.1	0.93–1.25	12.0–16.1

			Male and Female	
Test	Age	n	ng/mL	pmol/L
4	0–3 d	**	2.0–5.0	25.7–64.3
	3–30 d		0.9–2.2	11.6–28.3
	31 d–18 y		0.8–2.0	10.3–25.7

Specimen Type(s)	1–3	Plasma/serum
	4	Serum
Reference(s)	1	Soldin SJ,. Morales A, Albalos F, et al. Pediatric reference ranges on the Abbott IMx for FSH, LH, prolactin, TSH, T_4, T_3, free T_4, free T_4, T-uptake, IgE and ferritin. Clin Biochem 1995;28:603–6.
	2	Murthy JN, Hicks, JM, Soldin SJ. Evaluation of the Technicon Immuno I Random Access Immunoassay Analyzer and calculation of pediatric reference ranges for endocrine tests, T-uptake, and ferritin. Clin Biochem 1995;28:181–5.
	3	Ghoshal AK, Soldin SJ. Evaluation of the Dade Behring Dimension RxL: integrated chemistry system-pediatric reference ranges. Clin Chim Acta 2003;331:135–46.
	4	Dugaw KA, Jack RM, Rutledge J. Pediatric reference ranges for TSH, free T4, total T4, total T3 and T3 uptake on the Vitros ECi analyzer. Chem Chem 2001;47:A108. (Abstract)
Method(s)	1	Abbott IMx Analyzer (Abbott Diagnostics, Inc., Abbott Park, IL).
	2	Immuno I (Bayer Corp., Tarrytown, NY).
	3	The FT_4 method for the Dimension® RxL system with the heterogeneous immunoassay module is a two-step enzyme immunoassay based on a sequential competitive format. Dade Behring Dimension RxL Analyzer (Dade Behring Inc., Newark, DE).
	4	Chemiluminescent immunoassay, Vitros ECi (Ortho-Clinical Diagnostics, Raritan, NJ).
Comment(s)	1,2	Study used hospitalized patients and a computerized approach adapted from the Hoffmann technique. Values are 2.5–97.5th percentiles.
	3	*Numbers not provided.
		Reference ranges were obtained by comparing results from previously published data and using regression equations.
	4	**Ages ranged from 1 h to 18 y with a total of 119 specimens.

TRANSFERRIN

	Male and Female			
Test	Age	n	mg/dL	g/L
1	0–5 d	73	124–388	1.24–3.88
	1–3 y	51	190–302	1.90–3.02
	4–6 y	39	181–329	1.81–3.29
	7–9 y	39	196–314	1.96–3.14
	10–13 y	110	195–385	1.95–3.85
	14–19 y	78	203–386	2.03–3.86

		Male			Female		
Test	Age	n	mg/dL	g/L	n	mg/dL	g/L
2	1–30 d	41	84–178	0.84–1.78	36	80–181	0.80–1.81
	31–182 d	88	92–283	0.92–2.83	70	111–269	1.11–2.69
	183–365 d	40	155–311	1.55–3.11	29	127–317	1.27–3.17
	1–3 y	157	171–318	1.71–3.18	142	130–332	1.30–3.32
	4–6 y	108	176–305	1.76–3.05	94	151–347	1.51–3.47
	7–9 y	91	130–307	1.30–3.07	91	162–320	1.62–3.20
	10–12 y	85	151–331	1.51–3.31	105	161–328	1.61–3.28
	13–15 y	82	149–325	1.49–3.25	84	168–340	1.68–3.40
	16–18 y	60	169–303	1.69–3.03	82	157–362	1.57–3.62

			Male		Female	
Test	Age	n	mg/dL	g/L	mg/dL	g/L
3	1–30 d	*	86–174	0.86–1.74	83–176	0.83–1.76
	31–182 d		94–271	0.94–2.71	111–258	1.11–2.58
	183–365 d		152–296	1.52–2.96	126–303	1.26–3.03
	1–3 y		167–304	1.67–3.04	129–317	1.29–3.17
	4–6y		172–291	1.72–2.91	149–331	1.49–3.31
	7–9 y		129–293	1.29–2.93	159–305	1.59–3.05
	10–12 y		115–316	1.15–3.16	158–313	1.58–3.13
	13–15 y		147–310	1.47–3.10	164–324	1.64–3.24
	16–18 y		165–289	1.65–2.89	154–344	1.54–3.44

Specimen Type(s)	1,2	Plasma/serum
	3	Serum
Reference(s)	The above values have been adjusted from those published in references 1 and 2 to convert results to current IFCC units using the new IFCC guidelines/standards.	
	1	Lockitch G, Halstead AC, Quigley G, et al. Age- and sex-specific pediatric reference intervals: study, design and methods illustrated by measurement of serum proteins with the Behring LN nephelometer. Clin Chem 1988;34:1618–21.
	2	Soldin SJ, Hicks JM, Bailey J, et al. Pediatric reference ranges for total iron binding capacity and transferrin. Clin Chem 1997;43:S200. (Abstract)
	3	Ghoshal AK, Soldin SJ. Evaluation of the Dade Behring Dimension RxL: integrated chemistry system-pediatric reference ranges. Clin Chim Acta 2003;331:135–46.
Method(s)	1	Nephelometric, Behring LN nephelometer (Behring Diagnostics, Hoechst, Inc., Montreal, Canada).
	2	Nephelometric, Behring LN nephelometer (Behring Diagnostics, Inc., Westwood, MA).
	3	The method is a quantitative turbidimetric assay using endpoint detection based on the precipitation of transferrin by its polyclonal antibody. Dade Behring Dimension RxL Analyzer (Dade Behring Inc., Newark, DE).
Comment(s)	1	Normal healthy children. Values provided are 2.5–97.5th percentiles.
	2	Study used hospitalized patients and a computerized adaptation of the Hoffmann technique. Results are 2.5–97.5th percentiles.
	3	*Numbers not provided.
		Reference ranges were obtained by comparing results from previously published data and using regression equations.

TRANSFERRIN SATURATION

	Male		Female	
Age	**n**		**n**	
1–5 y*	44	0.07–0.44	44	0.07–0.44
6–9 y*	50	0.17–0.42	50	0.17–0.42
10–14 y	31	0.11–0.36	40	0.02–0.40
14–19 y	65	0.06–0.33	110	0.06–0.33
Specimen Type(s)	Plasma			
Reference(s)	Lockitch G, Halstead AC, Wadsworth L, et al. Age- and sex-specific pediatric reference intervals and correlations for zinc, copper, selenium, iron, vitamins A and E, and related proteins. Clin Chem 1988;34:1625–8.			
Method(s)	Transferrin was measured by nephelometry. Behring LN with Behring kit (Behringwerke, Marburg, Germany).			
Comment(s)	The study population was healthy children. Non-parametric methods were used to determine the 0.025 and 0.975 fractiles. *Males and females were not studied separately. Transferrin saturation = iron (μmoL/L) ÷ TIBC.			

TRIGLYCERIDES

		Male			Female		
Test	Age	n	mmol/L	mg/dL	n	mmol/L	mg/dL
1	0–7 d	149	0.24–2.06	21–182	142	0.32–1.88	28–166
	8–30 d	283	0.34–2.08	30–184	172	0.34–1.86	30–165
	31–90 d	247	0.45–1.98	40–175	171	0.40–3.19	35–282
	91–180 d	132	0.51–3.29	45–291	126	0.57–4.01	50–355
	181–365 d	286	0.51–5.66	45–501	261	0.41–4.87	36–431
2	1–3 y	49*	0.31–1.41	27–125	49*	0.31–1.41	27–125
	4–6 y	38*	0.36–1.31	32–116	38*	0.36–1.31	32–116
	7–9 y	72*	0.32–1.46	28–129	72*	0.32–1.46	28–129
	10–11 y	28	0.27–1.55	24–137	34	0.44–1.58	39–140
	12–13 y	32	0.27–1.64	24–145	40	0.42–1.47	37–130
	14–15 y	39	0.38–1.86	34–165	50	0.43–1.52	38–135
	16–19 y	41	0.38–1.58	34–140	68	0.42–1.58	37–140
3	Cord blood serum arterial	397**	0.10–1.04	9–92	397**	0.10–1.04	9–92
	Cord blood serum venous	397**	0.13–0.97	12–86	397**	0.13–0.97	12–86
4	0–7 d	***	0.21–1.97	19–174	***	0.29–1.80	26–159
	8–30 d		0.42–3.15	37–279		0.37–3.05	33–270
	31–90 d		0.47–3.15	42–279		0.38–3.84	34–340
	1–3 y		0.28–1.34	25–119		0.28–1.34	25–119
	4–6 y		0.34–1.24	30–110		0.34–1.24	30–110
	7–9 y		0.29–1.39	26–123		0.29–1.39	26–123
	10–11 y		0.25–1.48	22–131		0.42–1.51	37–134
	12–13 y		0.25–1.56	22–138		0.40–1.40	35–124
	14–15 y		0.36–1.78	32–158		0.41–1.46	36–129
	16–19 y		0.36–1.51	32–134		0.40–1.51	35–134

Specimen Type(s)	1	Serum
	2–4	Plasma/serum
Reference(s)	1	Soldin SJ, Morse AS. Pediatric reference ranges for calcium and triglycerides in children <1 year old using the Vitros 500 Analyzer. Clin Chem 1998;44:A16. (Abstract)
	2	Lockitch G, Halstead AC, Albersheim S, et al. Age- and sex-specific pediatric reference intervals for biochemistry analytes as measured with the Ektachem-700 analyzer. Clin Chem 1988;34:1622–5.
	3	Perkins SL, Livesey JF, Belcher J. Reference intervals for 21 clinical chemistry analytes in arterial and venous umbilical cord blood. Clin Chem 1993;39:1041–4.
	4	Ghoshal AK, Soldin SJ. Evaluation of the Dade Behring Dimension RxL: integrated chemistry system-pediatric reference ranges. Clin Chim Acta 2003;331:135–46.
Method(s)	1,2	Glycerol phosphate oxidase on Vitros 700 analyzer (Ortho-Clinical Diagnostics, Raritan, NJ).
	3	Hitachi 737 with Boehringer Mannheim reagents (Boehringer Mannheim Canada, Montreal Canada).
	4	The sample is pre-incubated with lipase enzyme reagent, which converts triglycerides into free glycerol and fatty acids. The liberated glycerol is determined enzymatically using glycerol dehydrogenase and NAD. Dade Behring Dimension RxL Analyzer (Dade Behring Inc., Newark, DE).
Comment(s)	1	The study used plasma/serum obtained from hospitalized patients and employed Chauvenet's criteria to remove outliers and a computerized approach adapted from the Hoffmann technique to obtain the 2.5–97.5th percentiles. Fasting samples were not obtained in these neonates and infants, which accounts for the somewhat elevated 97.5th percentiles.
	2	A healthy population of children 1–19 y was studied. Results are 2.5–97.5th percentiles.
		*No significant differences were found for males and females. These ranges were therefore derived from combined data.
	3	Results are 2.5–97.5th percentiles.
		**No significant differences were found for males and females. These ranges were therefore derived from combined data.
	4	***Numbers not provided.
		Reference ranges were obtained by comparing results from previously published data and using regression equations.

TRIIODOTHYRONINE (T_3)

		Male			Female		
Test	Age	n	ng/dL	nmol/L	n	ng/dL	nmol/L
1	1–30 d	50	15–210	0.2–3.2	47	15–200	0.2–3.1
	1–12 mo	111	95–275	1.5–4.2	100	50–264	0.8–4.1
	1–5 y	101	80–253	1.2–3.9	115	126–258	1.9–4.0
	6–10 y	99	96–232	1.5–3.6	99	104–227	1.6–3.5
	11–15 y	97	73–199	1.1–3.1	100	96–211	1.5–3.2
	16–18 y	100	69–201	1.1–3.1	104	91–164	1.4–2.5
2	1–30 d	227	29–179	0.4–2.7	173	34–173	0.5–2.7
	1–3 y	115	58–190	0.9–2.9	127	54–199	0.8–3.1
	4–6 y	112	59–177	0.9–2.7	105	58–173	0.9–2.7
	7–12 y	114	52–192	0.8–2.9	110	53–171	0.8–2.6
	13–18 y	113	64–157	1.0–2.4	121	40–156	0.6–2.4
3	1–30 d	50	51–184	0.7–2.9	47	48–177	0.7–2.8
	1–12 mo	111	103–229	1.6–3.5	100	73–221	1.1–3.4
	1–5 y	101	93–213	1.4–3.3	115	126–216	1.9–3.4
	6–10 y	99	104–198	1.6–3.0	99	110–195	1.7–3.0
	11–15 y	97	88–176	1.3–2.7	100	104–184	1.6–2.9
	16–18 y	100	86–178	1.3–2.8	104	101–151	1.5–2.4
4	1–30 d	*	40–180	0.6–2.8	*	40–173	0.6–2.7
	1–12 mo		94–228	1.4–3.5		62–220	1.0–3.4
	1–5 y		83–211	1.3–3.2		118–214	1.0–3.3
	6–10 y		95–195	1.5–3.0		101–192	1.6–2.9
	11–15 y		78–172	1.2–2.6		95–180	1.5–2.8
	16–18 y		76–174	1.2–2.7		92–145	1.4–2.2

		Male and Female		
Test	Age	n	ng/dL	nmol/L
5	0–3 d	**	60–300	0.9–4.7
	4–365 d		90–260	1.4–4.0
	1–6 y		90–240	1.4–3.7
	7–11 y		90–230	1.4–3.6
	12–18 y		100–210	1.5–3.3

Specimen Type(s)	1–4	Plasma/serum
	5	Serum
Reference(s)	1	Soldin SJ, Morales A, Albalos F, et al. Pediatric reference ranges on the Abbott IMx for FSH, LH, prolactin, TSH, T_4, T_3, free T_4, free T_3, T-uptake, IgE, and ferritin. Clin Biochem 1995;28:603–6.
	2	Soldin SJ, Hicks JM, Bailey J, et al. Pediatric reference ranges for triiodothyronine. Clin Chem 1997;43:S199. (Abstract)
	3	Murthy JN, Hicks JM, Soldin SJ. Evaluation of the Technicon Immuno I Random Access Immunoassay Analyzer and calculation of pediatric reference ranges for endocrine tests, T-uptake, and ferritin. Clin Biochem 1995;28:181–5.
	4	DPC IMMULITE® 1000 reference ranges used at Children's National Medical Center, Washington, DC.
	5	Dugaw KA, Jack RM, Rutledge J. Pediatric reference ranges for TSH, free T4, total T4, total T3 and T3 uptake on the Vitros ECi analyzer. Chem Chem 2001;47:A108. (Abstract)
Method(s)	1	Abbott IMx Analyzer (Abbott Diagnostics, Inc., Abbott Park, IL).
	2	T_3 (Diagnostic Products, Inc. Los Angeles, CA).
	3	Immuno I (Bayer Corp., Tarrytown NY).
	4	DPC IMMULITE® 1000 analyzer (Diagnostic Products Corporation, Los Angeles, CA).
	5	Chemiluminescent immunoassay, Vitros ECi (Ortho-Clinical Diagnostics, Raritan, NJ).
Comment(s)	1–3	Study used hospitalized patients and a computerized approach adapted from the Hoffmann technique. Values are 2.5–97.5th percentiles.
	4	*Numbers not provided.
		Pediatric reference ranges were determined by using regression equations to transform the previously established ranges on the Bayer Immuno 1 (Bayer Corp., Tarrytown, NY) based on the Hoffmann technique to their corresponding values on the IMMULITE® 1000.
	5	**Ages ranged from 1 h to 18 y with a total of 119 specimens.

TRIIODOTHYRONINE, FREE (FREE T_3)

		Male			Female		
Test	**Age**	**n**	**ng/dL**	**pmol/L**	**n**	**ng/dL**	**pmol/L**
1	1–3 d	24	0.14–0.48	2.2–7.4	26	0.14–0.54	2.2–8.3
	4–30 d	73	0.14–0.55	2.2–8.4	62	0.15–0.50	2.3–7.7
	1–12 mo	52	0.20–0.69	3.1–10.6	52	0.25–0.65	3.8–10.0
	1–5 y	100	0.24–0.67	3.7–10.3	99	0.30–0.60	4.6–9.2
	6–10 y	104	0.29–0.60	4.4–9.2	101	0.27–0.62	4.1–9.5
	11–15 y	102	0.31–0.59	4.8–9.1	102	0.26–0.57	4.0–8.8
	16–18 y	101	0.35–0.57	5.4–8.8	98	0.28–0.52	4.3–8.0
2	0–6 y	62	0.40–0.71	6.1–10.9	62	0.40–0.71	6.1–10.9
	7–12 y	43	0.33–0.80	5.1–12.3	43	0.33–0.80	5.1–12.3
	13–18 y	33	0.23–0.75	3.5–11.5	33	0.23–0.75	3.5–11.5

Specimen Type(s)	1,2	Plasma/serum
Reference(s)	1	Soldin SJ, Morales A, Albalos F, et al. Pediatric reference ranges on the Abbott IMx for FSH, LH, prolactin, TSH, T_4, T_3, free T_4, free T_3, T-uptake, IgE, and ferritin. Clin Biochem 1995;28:603–6.
	2	Butler J, Moore P, Mieli-Vergani G, et al. Serum free thyroxine and free triiodothyronine in normal children. Ann Clin Biochem 1988;25:536–9.
Method(s)	1	Abbott IMx Analyzer (Abbott Diagnostics, Inc., Abbott Park, IL).
	2	RIA Amerlex (Amersham International plc, Cardiff, UK).
Comment(s)	1	Study used hospitalized patients and a computerized approach adapted from the Hoffmann technique. Values are 2.5–97.5th percentiles.
	2	Normal school children were used in this study. Data were log-transformed and the mean and standard deviation were calculated.

TRIIODOTHYRONINE UPTAKE TEST (T_3U)

		Male		Female	
Test	**Age**	**n**	**%**	**n**	**%**
1	1–365 d	96	23–34	95	23–36
	1–3 y	147	24–35	127	24–36
	4–6 y	124	24–34	116	24–35
	7–12 y	127	24–33	136	22–35
	13–15 y	142	25–37	111	23–37
	16–18 y	47	24–38	114	23–35
		Male		**Female**	
Test	**Age**	**n**	**T-Uptake Units**	**n**	**T-Uptake Units**
2	1–30 d	34	0.26–1.66	44	0.41–1.44
	1 mo–5 y	99	0.72–1.35	141	0.81–1.36
	6–10 y	96	0.74–1.22	98	0.78–1.18
	11–19 y	113	0.58–1.23	108	0.76–1.25
		Male		**Female**	
Test	**Age**	**n**	**Ratio**	**n**	**Ratio**
3	0–1 mo	34	0.71–1.25	44	0.88–1.08
	1 mo–5 y	99	0.89–1.13	141	0.92–1.13
	6–10 y	96	0.84–1.08	98	0.82–1.10
	11–18 y	113	0.88–1.08	108	0.77–1.16
		Male and Female			
Test	**Age**	**n**	**%**		
4	0–30 d	*	24–48		
	31 d–18 y		26–39		

Specimen Type(s)	1–3	Plasma/serum
	4	Serum
Reference(s)	1	Soldin SJ, Cook J, Beatey J, et al. Pediatric reference ranges for thyroxine and triiodothyronine. Clin Chem 1992;38:960. (Abstract)
	2	Soldin SJ, Morales A, Albalos F, et al. Pediatric reference ranges on the Abbott IMx for FSH, LH, prolactin, TSH, T_4, T_3, free T_4, free T_3, T-uptake, IgE, and ferritin. Clin Biochem 1995;28:603–6.
	3	Murthy JN, Hicks JM, Soldin SJ. Evaluation of the Technicon Immuno I Random Access Immunoassay Analyzer and calculation of pediatric reference ranges for endocrine tests, T-uptake, and ferritin. Clin Biochem 1995;28:181–5.
	4	Dugaw KA, Jack RM, Rutledge J. Pediatric reference ranges for TSH, free T4, total T4, total T3 and T3 uptake on the Vitros ECi analyzer. Chem Chem 2001;47:A108. (Abstract)
Method(s)	1	RIA Quantimmune® 11 (Bio-Rad, Bio-Rad Laboratories, Richmond, CA).
	2	Abbott IMx (Abbott Laboratories, Abbott Park, IL).
	3	Immuno I (Bayer Corp., Tarrytown, NY).
	4	Chemiluminescent immunoassay, Vitros ECi (Ortho-Clinical Diagnostics, Raritan, NJ).
Comment(s)	1–3	Study used hospitalized patients and a computerized approach adapted from the Hoffmann technique to obtain the 2.5–97.5th percentiles.
	4	*Ages ranged from 1 h to 18 y with a total of 119 specimens.

TROPONIN I

Male and Female			
Test	**Age**	**n**	**97.5th percentile μg/L**
1	0–30 d	97	<4.8
	31–90 d	46	<0.4
	3–6 mo	91	<0.3
	7–12 mo	53	<0.2
	1–18 y	57	<0.1
2	0–30 d	*	<8.4
	31–90 d		<0.7
	3–6 mo		<0.5
	7–12 mo		<0.3
	1–18 y		<0.1

Specimen Type(s)	1,2	Plasma/serum
Reference(s)	1	Soldin SJ, Murthy JN, Agarwalla PK, et al. Pediatric reference ranges for creatine kinase, CKMB, troponin I, and cortisol. Clin Biochem 1999;32:77–80.
	2	Ghoshal AK, Soldin SJ. Evaluation of the Dade Behring Dimension RxL: integrated chemistry system-pediatric reference ranges. Clin Chim Acta 2003;331:135–46.
Method(s)	1	Bayer Immuno I with Bayer reagents (Bayer Corp., Tarrytown, NY).
	2	The TROP method is a one-step enzyme immunoassay based on the "sandwich" principle. Dade Behring Dimension RxL Analyzer (Dade Behring Inc., Newark, DE).
Comment(s)	1	Studies used hospitalized patients and a computerized approach to removing outliers adapted from the Hoffmann technique. Values are the 97.5th percentiles.
	2	*Numbers not provided.
		Reference ranges were obtained by comparing results from previously published data and using regression equations.

UREA NITROGEN

		Male			Female		
Test	**Age**	**n**	**mg/dL**	**mmol/L**	**n**	**mg/dL**	**mmol/L**
1	1–7 d	171	2–13	0.7–4.6	114	2–13	0.7–4.6
	8–30 d	209	2–16	0.7–5.7	154	2–15	0.7–5.4
	1–3 mo	278	2–12	0.7–4.3	274	2–14	0.7–5.0
	4–6 mo	144	1–14	0.4–5.0	139	1–13	0.4–4.6
	7–12 mo	204	2–14	0.7–5.0	160	1–13	0.4–4.6
2	1–30 d	51	4–12	1.4–4.3	43	3–17	1.1–6.1
	1–12 mo	69	2–13	0.7–4.6	60	4–14	1.4–5.0
	1–3 y	104	3–12	1.1–4.3	127	3–14	1.1–5.0
	4–6 y	140	3–16	1.1–5.7	122	4–14	1.4–5.0
	7–9 y	124	4–16	1.4–5.7	121	4–16	1.4–5.7
	10–12 y	133	5–18	1.8–6.4	125	5–16	1.8–5.7
	13–15 y	141	7–18	2.5–6.4	153	4–15	1.4–5.4
	16–18 y	111	5–20	1.8–7.1	120	4–15	1.4–5.4

		Male and Female		
Test	**Age**	**n**	**mg/dL**	**mmol/L**
3	1–3 y	50	5–17	1.8–6.0
	4–6 y	38	7–17	2.5–6.0
	7–9 y	72	7–17	2.5–6.0
	10–11 y	62	7–17	2.5–6.0
	12–13 y	73	7–17	2.5–6.0
	14–15 y	91	8–21	2.9–7.5
	16–19 y	107	8–21	2.9–7.5
4	1–7 d	*	1–13	0.3–3.5
	8–30 d		1–16	0.3–4.3
	1–3 mo		1–12	0.3–3.2
	4–12 mo		1–14	0.3–3.8
	1–3 y		4–17	1.1–4.6
	4–13 y		6–17	1.6–4.6
	14–19 y		7–21	1.9–5.7

Specimen Type(s)	1,2,4	Plasma/serum
	3	Plasma
Reference(s)	1	Soldin SJ, Savwoir TV, Guo Y. Pediatric reference ranges for gamma-glutamyltransferase and urea nitrogen during the first year of life on the Vitros 500 analyzer. Clin Chem 1997;43:S199. (Abstract)
	2	Soldin SJ, Bailey J, Beatey J, et al. Pediatric reference ranges for blood urea nitrogen (BUN) on the Hitachi 747 analyzer. Clin Chem 1996;42:S307. (Abstract)
	3	Lockitch G, Halsted AC, Albersheim S, et al. Age- and sex-specific pediatric reference intervals for biochemistry analytes as measured with the Ektachem-700 analyzer. Clin Chem 1988;34:1622–5. (Abstract)
	4	Ghoshal AK, Soldin SJ. Evaluation of the Dade Behring Dimension RxL: integrated chemistry system-pediatric reference ranges. Clin Chim Acta 2003;331:135–46.
Method(s)	1,3	Urease method, Vitros 500 & 700 (Ortho-Clinical Diagnostics, Raritan, NJ).
	2	Hitachi 747 with Boehringer Mannheim reagents (Boehringer Mannheim Diagnostics, Indianapolis, IN).
	4	The urea nitrogen method employs a urease/glutamate dehydrogenase-coupled enzymatic technique. Dade Behring Dimension RxL Analyzer (Dade Behring Inc., Newark, DE).
Comment(s)	1,2	Study used hospitalized patients and a computerized approach adapted from the Hoffmann technique to obtain 2.5–97.5th percentiles.
	3	Study used normal healthy children. Values are 2.5–97th percentiles.
	4	*Numbers not provided.
		Reference ranges were obtained by comparing results from previously published data and using regression equations.

URIC ACID

		Male			Female		
Test	Age	n	mg/dL	µmol/L	n	mg/dL	µmol/L
1	1–30 d	72	1.3–4.9	80–290	47	1.4–6.2	80–370
	1–3 mo	83	1.4–5.3	80–310	64	1.4–5.8	80–340
	4–6 mo	104	1.5–6.3	90–370	52	1.4–6.2	80–370
	7–12 mo	97	1.5–6.6	90–390	65	1.5–6.2	90–370
2	1–3 y	49*	1.8–5.0	105–300	49*	1.8–5.0	105–300
	4–6 y	38*	2.2–4.7	130–280	38*	2.2–4.7	130–280
	7–9 y	72*	2.0–5.0	120–295	72*	2.0–5.0	120–295
	10–11 y	28	2.3–5.4	135–320	34	3.0–4.7	180–280
	12–13 y	32	2.7–6.7	160–400	40	3.0–5.8	180–345
	14–15 y	39	2.4–7.8	140–465	50	3.0–5.8	180–345
	16–19 y	41	4.0–8.6	235–510	68	3.0–5.9	180–350
3	1–30 d	84	1.2–3.9	71–230	91	1.0–4.6	59–271
	31–365 d	138	1.2–5.6	71–330	110	1.1–5.4	65–319
	1–3 y	149	2.1–5.6	124–330	105	1.8–5.0	106–295
	4–6 y	120	1.8–5.5	106–325	91	2.0–5.1	118–301
	7–9 y	123	1.8–5.4	106–319	115	1.8–5.5	106–325
	10–12 y	121	2.2–5.8	130–342	92	2.5–5.9	148–348
	13–15 y	106	3.1–7.0	183–413	116	2.2–6.4	130–378
	16–18 y	82	2.1–7.6	124–448	111	2.4–6.6	142–389
4	1–30 d	**	1.2–4.9	71–291	**	1.3–6.2	77–369
	1–3 mo		1.3–5.3	77–315		1.3–5.8	77–345
	4–6 mo		1.4–6.4	83–381		1.3–6.2	77–369
	7–12 mo		1.4–6.7	83–399		1.4–6.2	83–369
	1–3 y		1.7–5.0	101–297		1.7–5.0	101–297
	4–6 y		2.2–4.7	131–280		2.2–4.7	131–280
	7–9 y		1.9–5.0	113–297		1.9–5.0	113–297
	10–11 y		2.3–5.4	137–321		3.0–4.7	178–280
	12–13 y		2.7–6.8	161–404		3.0–5.8	178–345
	14–15 y		2.4–7.9	143–470		3.0–5.8	178–345
	16–19 y		4.0–8.7	238–518		3.0–5.9	178–351

Specimen Type(s)	1,3,4	Plasma/serum
	2	Serum
Reference(s)	1	Soldin SJ, Savwoir TV, Guo Y. Pediatric reference ranges for lactate dehydrogenase and uric acid during the first year of life on the Vitros 500 analyzer. Clin Chem 1997;43:S199. (Abstract)
	2	Lockitch G, Halstead AC, Albersheim S, et al. Age- and sex-specific pediatric reference intervals for biochemistry analytes as measured with the Ektachem-700 analyzer. Clin Chem 1988;34:1622–5.
	3	Soldin SJ, Bailey J, Beatey J, et al. Pediatric reference ranges for uric acid. Clin Chem 1996;42:S308. (Abstract)
	4	Ghoshal AK, Soldin SJ. Evaluation of the Dade Behring Dimension RxL: integrated chemistry system-pediatric reference ranges. Clin Chim Acta 2003;331:135–46.
Method(s)	1,2	Uricase on Vitros analyzer (Ortho-Clinical Diagnostics, Raritan, NJ).
	3	Hitachi 747 analyzer with Boehringer Mannheim reagents (Boehringer Mannheim Diagnostics, Indianapolis, IN).
	4	Measurement of uric acid by monitoring the loss of absorbance at 293 nm following uricase treatment is generally recognized as being more specific and less subject to interference than other indirect methods. Principles of procedure: Uric acid, which absorbs light at 293 nm, is converted by uricase to allantoin, which is nonabsorbing at 293 nm. The change in absorbance at 293 nm due to the disappearance of uric acid is directly proportional to the concentration of uric acid in the sample and is measured using a bichromatic (293,700 nm) endpoint technique. Dade Behring Dimension RxL Analyzer (Dade Behring Inc., Newark, DE).
Comment(s)	1,3	Study used hospitalized patients and a computerized approach adapted from the Hoffmann technique to obtain the 2.5–97.5th percentiles.
	2	Study performed on healthy children. Results are 2.5–97.5th percentiles.
		*No significant differences were found for males and females. These ranges were therefore derived from combined data.
	4	**Numbers not provided.
		Reference ranges were obtained by comparing results from previously published data and using regression equations.

URINE VOLUME (24 h)

Male and Female		
Age	**mL**	**L**
Newborn	50–300	0.05–0.30
Infant	350–550	0.35–0.55
Child	500–1000	0.50–1.00
Adolescent	700–1400	0.70–1.40
Thereafter	600–1800	0.60–1.80

Specimen Type(s)	Urine (24 h)
Reference	Behrman RE, ed. Nelson textbook of pediatrics, 14th ed. Philadelphia, PA: WB Saunders Company, 1992:1824.
Method(s)	Not given.
Comments	Varies with intake and other factors.

VANILLYLMANDELIC ACID (VMA) (URINE),4–HYDROXY–3–METHOXYMANDELIC ACID

Male and Female							
Test	**Age**	**n**	**mg/g creatinine**	**mmol/ mol creatinine**	**n**	**mg/24h**	**µmol/24h**
1	0–1 y	37	<18.8	<10.7	48	<2.3	<11.6
	2–4 y	49	<11.0	<6.3	34	<3.0	<15.1
	5–9 y	79	<8.3	<4.7	20	<3.5	<17.7
	10–19 y	55	<8.2	<4.7	40	<6.0	<30.3
	>19 y	56	<6.0	<3.4	56	<6.8	<34.3
2	0–3 mo	12	5.9–37.0	3.4–21.1			
	3–12 mo	28	8.4–43.8	4.8–25.0			
	1–2 y	15	7.9–23.0	4.5–13.1			
	2–5 y	22	2.9–23.0	1.7–13.1			
	5–10 y	21	5.8–18.7	3.3–10.7			
	10–15 y	13	1.6–10.6	0.9–6.1			
	>15 y	8	2.8–8.3	1.6–4.7			

Specimen Type	1,2	Urine
Reference(s)	1	Soldin SJ, Hill JG. Liquid chromatographic analysis for urinary 4-hydroxy-3-methoxy-mandelic acid and 4-hydroxy-3-methoxyphenyl acetic acid and its use in investigation of neural crest tumors. Clin Chem 1981;27:502–3.
	2	Tuchman M, Morris CL, Ramnaraine ML, et al. Value of random urinary homovanillic acid and vanillylmandelic acid levels in the diagnosis and management of patients with neuroblastoma: comparison with 24-h urine collections. Pediatrics 1985;75:324–8.
Method(s)	1	HPLC with electrochemical detection.
	2	Capillary gas chromatography.
Comment(s)	1	Analysis performed on patients under investigation of hypertension or not suspected of having a neural crest tumor. All patients studied free of neoplasia. Results are 95th percentile.
	2	Values are 0–100th percentiles. Normal values were obtained from 93 pediatric patients in whom the diagnosis of neural crest tumors had been excluded.

VITAMIN A (RETINOL)

Male and Female				
Test	**Age**	**n**	**µmol/L**	**µg/dL**
1	At birth	25	0.49–1.81	14–52
	Adult (mothers)	25	0.71–2.27	20–65
2	1–6 y	62*	0.7–1.5	20–43
	7–12 y	23*	0.9–1.7	20–49
	13–19 y	24*	0.9–2.5	26–72
3	Term neonates (>37 weeks)	41	0.63–1.75	18–50
	Preterm neonates (<36.6 weeks)	58	0.46–1.61	13–46

Specimen Type	1,2	Serum
	3	Plasma/serum
Reference(s)	1	Jansson L, Nilsson B. Serum retinol and retinal-binding protein in mothers and infants at delivery. Biol Neonate 1983;43:269–71.
	2	Lockitch G, Halstead A, Wadsworth L, et al. Age- and sex-specific pediatric reference intervals for zinc, copper, selenium, iron, vitamins A and E, and related proteins. Clin Chem 1988;34:1625–8.
	3	Cardona-Pérez A, et al. Cord blood retinol and retinol binding protein in preterm and term neonates. Nutrition Research 1996:16;191–6.
Method(s)	1	Affinity chromatography and HPLC. See reference.
	2,3	HPLC. See reference for description.
Comments	1	Results are mean ±2 SDs. Retinol studied in 25 healthy newborns and their mothers.
	2	The study population was healthy children. Non-parametric methods were used to determine the 0.025 and 0.975 fractiles.
		*No significant differences were found for males and females. These ranges were therefore derived from combined data.
	3	Healthy term and preterm neonates studied. Results are 2.5–97.5th percentiles.

VITAMIN B_{12}

		Male		Female	
Test	Age	n	pmol/L	n	pmol/L
1	0–1 y	127	216–891	94	168–1117
	2–3 y	142	195–897	133	307–892
	4–6 y	156	181–795	111	231–1038
	7–9 y	103	200–863	103	182–866
	10–12 y	105	135–803	94	145–752
	13–18 y	159	158–638	159	134–605
2	Newborn[a]	*	130–590	*	130–590
	After newborn period[a]		100–520		100–520
	Adult[b]		120–700		120–700

Specimen Type(s)	1	Plasma
	2	Serum
Reference(s)	1	Hicks JM, Cook J, Godwin ID, et al. Vitamin B_{12} and folate: Pediatric reference ranges. Arch Pathol Lab Med 1993;117:704–6.
	2	[a]Behrman RE, Vaughan VC, eds. Nelson textbook of pediatrics. Philadelphia, PA: WB Saunders 1983:1827–60.
		[b]Hall CA, Bardwell SA, Allen ES, et al. Variation in plasma folate levels among groups of healthy persons. Am J Clin Nutr, 1975;28:854–7.
Method(s)	1	Radioimmunoassay Quantaphase B_{12} (Bio-Rad, Hercules, CA).
	2	See references for particulars of assays used.
Comment(s)	1	Study used hospitalized patients and a computerized approach adapted from the Hoffmann technique. Values are 2.5–97.5th percentiles.
	2	*Numbers not available.

25–HYDROXY VITAMIN D (25 OH VIT D)

		Male				Female			
		Summer		Winter		Summer		Winter	
Test	Age	n	μg/L	n	μg/L	n	μg/L	n	μg/L
1	1–30 d	61	6.2–33.4	46	3.3–28.8	62	6.4–30.7	34	1.9–32.0
	31–365 d	76	18.6–36.9	71	7.4–53.3	87	16.0–48.2	64	11.6–36.7
	1–3 y	144	6.9–46.8	90	10.1–39.0	129	11.6–48.9	109	11.3–41.1
	4–12 y	171	4.6–37.4	151	6.3–36.0	161	2.8–36.7	59	8.4–29.9
	13–18 y	121	4.3–31.4	41	2.0–29.1	130	2.3–28.3	63	1.8–20.4
2	1–30 d	61	11–44	46	8–38	62	11–41	34	6–42
	1 mo–<1 y	76	26–48	71	13–68	87	23–62	64	18–48
	1–3 y	144	12–60	90	16–51	129	18–62	109	17–53
	4–12 y	171	9–49	151	11–47	161	7–48	59	14–40
	13–18 y	121	9–41	41	6–39	130	7–38	63	6–28

Specimen Type(s)	1,2	Serum
Reference(s)	1	Soldin SJ, Hicks JM, Bailey J, et al. Pediatric reference ranges for 25 hydroxy vitamin D during the summer and winter. Clin Chem 1997;43:S200. (Abstract)
	2	Soldin SJ, Murthy J, Lauber B, et al. Pediatric reference ranges for 25-hydroxy vitamin D using DiaSorin kit. Clin Chem 1998;44;A14. (Abstract)
Method(s)	1	Radioimmunoassay. Nichols Institute Diagnostics Kit (Nichols Institue Diagnostics, Capistrano, CA).
	2	Radioimmunoassay, DiaSorin Kit (DiaSorin, Stillwater, MN).
Comment(s)	1,2	Study used hospitalized patients and a computerized approach adapted from the Hoffmann technique. Values are the 2.5–97.5th percentiles.

VITAMIN E (α-TOCOPHEROL)

Male and Female				
Test	Age	n	µmol/L	µg/mL
1	1–6 y	62*	7–21	3.0–9.0
	7–12 y	23*	10–21	4.0–9.0
	13–19 y	24*	13–24	6.0–10.1
2	Premature	**	1–8	0.5–3.5
	Full term		2–8	1.0–3.5
	2–5 mo		5–14	2.0–6.0
	6–24 mo		8–19	3.5–8.0
	2–12 y		13–21	5.5–9.0

Specimen Type(s)	1	Serum
	2	Plasma/serum
Reference(s)	1	Lockitch G, Halstead A, Wadsworth L, et al. Age- and sex-specific pediatric reference intervals for zinc, copper, selenium, iron, vitamins A and E, and related proteins. Clin Chem 1988;34:1625–8.
	2	Meites S, ed. Pediatric clinical chemistry, 3rd ed. Washington, DC: AACC Press, 1989:295–6.
Method(s)	1	HPLC.
	2	Hansen LG, Warwick WJ. An improved assay method for vitamins A and E using fluorometry. Am J Clin Path 1978;70:922–3.
		McWhirter WR. Plasma tocopherol in infants and children. Acta Path Scand 1975:64;446–8.
Comment(s)	1	The study population was healthy children. Non-parametric methods were used to determine the 0.025 and 0.975 fractiles.
		*No significant differences were found for males and females. These ranges were therefore derived from combined data.
	2	**See references for numbers used in study.

ZINC

		Male			Female		
Test	Age	n	µg/dL	µmol/L	n	µg/dL	µmol/L
1	0–5 d	27*	65–140	9.9–21.4	27*	65–140	9.9–21.4
	1–5 y	77*	67–118	10.3–18.1	77*	67–118	10.3–18.1
	6–9 y	44*	77–107	11.8–16.4	44*	77–107	11.8–16.4
	10–14 y	36	76–101	11.6–15.4	23	79–118	12.1–18.0
	15–19 y	55	64–104	9.8–15.4	31	60–101	9.2–15.4
2	3–5 y	38**	71–149	10.9–22.8	38**	71–149	10.9–22.8
	6–8 y	43**	70–128	10.7–19.6	43**	70–128	10.7–19.6
	9–11 y	52**	64–124	9.8–19.0	52**	64–124	9.8–19.0
	12–13 y	24**	45–125	6.9–19.1	24**	45–125	6.9–19.1
	14–16 y	29**	42–125	6.4–19.1	29**	42–125	6.4–19.1

		Male and Female		
Test	Age	n	µg/dL	µmol/L
3	0–<0.5 y	13	26–141	4.0–21.6
	0.5–<1.0 y	18	29–131	4.5–20.1
	1.0–<2.0 y	15	31–120	4.8–18.4
	2.0–<4.0 y	23	29–115	4.4–17.6
	4.0–<6.0 y	19	48–119	7.4–18.2
	6.0–<10.0 y	25	48–129	7.3–19.7
	10.0–<14.0 y	21	25–148	3.9–22.7
	14.0–<18.0 y	17	46–130	7.1–19.9

Specimen Type(s)	1,2	Serum
	3	Plasma/serum
Reference(s)	1	Lockitch G, Halstead A, Wadsworth L, et al. Age- and sex-specific pediatric reference intervals for zinc, copper, selenium, iron, vitamins A and E, and related proteins. Clin Chem 1988;34:1625–8.
	2	Malvy DJ-M, Arnaud J, Burtschy, B, et al. Reference values for serum, zinc and selenium of French healthy children. Eur J Epidemiol 1993;9:155–61.
	3	Rükgauer M, Klein J, Kruse–Jarres JD. Reference values for the trace elements copper, manganese, selenium, and zinc in the serum/plasma of children, Adolescents and Adults. J. Trace Elements Med Biol 1997;11:92–8.
Method(s)	1	Atomic absorption spectrometry with deuterium background correction Varian AA–1475 (Varian Canada, Inc., Georgetown, Canada).
	2	Perkin Elmer Atomic Absorption Spectrophotometry (Perkin Elmer Corporation, Rockville, MD).
	3	Atomic Absorption Spectrophotometry with Zeeman background compensation. Perkin Elmer ETAAS, Zeeman 3030 (Uberlingen, Germany).
Comment(s)	1	The study population was healthy children. Non-parametric methods were used to determine the 0.025 and 0.975 fractiles.
		*No significant differences were found for males and females. These ranges were therefore derived from combined data.
	2	The study population consisted of healthy French children. Values reported are the 2.5–97.5th percentiles.
		**No significant differences were found for males and females. These ranges were therefore derived from combined data.
	3	Study population was drawn from patients visiting the outpatient department or surgical or orthopedic ward for preoperative work up. Results are mean ±2 SDs (2.5th–97.5th percentiles).

ZINC PROTOPORPHYRIN

	Male			Female		
Age	**n**	**µg/dL**	**µmol/mol heme**	**n**	**µg/dL**	**µmol/mol heme**
0–12 mo	145	8.5–34.5	15.6–63.5	203	9–40	16.6–73.6
13–24 mo	695	10–34	18.4–62.6	725	11–32	20.2–58.9
2–5 y	1926	5–35	9.2–64.4	1822	10–31	18.4–57.0
6–9 y	522	6–31	11.0–57.0	408	9–30	16.6–55.2
10–17 y	61	5.5–31.5	10.1–58.0	61	5–33.5	9.2–61.6

Specimen Type(s)	50 µL anticoagulated whole blood containing either heparin or EDTA
Reference(s)	Soldin OP, Miller M, Soldin SJ. Pediatric reference ranges for zinc protoporphyrin. Clin Biochem 2003;36:21–5.
Method(s)	ZP Hematofluorometer Model 206 (AVIV Biomedical Inc., Lakewood, NJ).
Comment(s)	Study used hospitalized patients and a computerized approach to removing outliers. Values are 2.5–97.5th percentiles.

HEMATOLOGY TESTS

ATYPICAL LYMPHOCYTE COUNT (RELATIVE)

		Male		Female	
Test	Age	n	%	n	%
1	1–14 d	35	4.6–10.3	27	3.3–10.7
	15–30 d	33	3.9–12.2	27	3.9–9.7
	31–60 d	146	4.6–10.1	108	4.0–7.9
	61–180 d	458	3.7–10.2	3449	3.7–9.2
	0.5–<2 y	1524	3.4–8.7	1229	3.4–9.5
	2–<6 y	1600	2.6–7.3	1276	2.6–7.4
	6–<12 y	1453	2.1–5.6	1369	2.1–5.7
	12–<18 y	1375	1.9–5.3	1729	1.9–4.9
	>18 y	688	1.8–5.0	1058	1.8–4.6

Specimen Type(s)	Whole blood (K_2/EDTA anticoagulant)
Reference(s)	Brugnara C. Unpublished data.
Method(s)	Bayer ADVIA 120 (Bayer Diagnostics, Tarrytown, NY).
Comment(s)	Data from outpatients. A computerized approach adapted from the Hoffmann technique was used to remove outliers and obtain the 2.5th and 97.5th percentiles.

BASOPHIL COUNT (RELATIVE)

		Male			Female		
Test	Age	n	%	Number Fraction	n	%	Number Fraction
1	1–3 d	48	0–2.3	0–.023	50	0–2.1	0–.021
	4–7 d	33	0–2.4	0–.024	29	0–2.7	0–.027
	8–14 d	43	0–1.8	0–.018	58	0–1.8	0–.018
	15–30 d	137	0–1.2	0–.012	98	0–1.1	0–.011
	31–60 d	462	0–1.0	0–.010	359	0–0.9	0–.009
	61–180 d	1472	0–1.0	0–.010	1165	0–1.0	0–.010
	0.5–<2 y	5932	0–1.0	0–.010	4606	0–1.1	0–.011
	2–<6 y	6573	0–1.0	0–.010	5671	0–1.0	0–.010
	6–<12 y	5347	0–1.0	0–.010	4723	0–1.1	0–.011
	12–<18 y	3955	0–1.1	0–.011	5111	0–1.0	0–.010
	>18 y	1968	0–1.2	0–.012	3055	0–1.0	0–.010
2	1–14 d	35	0–1.8	0–.018	28	0–2.1	0–.021
	15–30 d	37	0–0.8	0–.008	32	0–1.1	0–.011
	31–60 d	151	0–0.9	0–.009	117	0–1.3	0–.013
	61–180 d	474	0–1.1	0–.011	365	0–1.1	0–.011
	0.5–<2 y	1570	0–1.0	0–.010	1262	0–1.1	0–.011
	2–<6 y	1635	0–0.9	0–.009	1310	0–0.9	0–.009
	6–<12 y	1467	0–0.8	0–.008	1391	0–0.9	0–.009
	12–<18 y	1391	0–0.8	0–.008	1755	0–0.8	0–.008
	>18 y	691	0–0.9	0–.009	1066	0–0.8	0–.008

Specimen Type(s)	1	Whole blood (K_3/EDTA anticoagulant)
	2	Whole blood (K_2/EDTA anticoagulant)
Reference(s)	1	Soldin SJ, Brugnara C, et al., eds. Pediatric reference ranges, 2nd ed. Washington, DC: AACC Press, 1997.
	2	Brugnara C. Unpublished data.
Method(s)	1	Bayer H*3 (Bayer Diagnostics, Tarrytown, NY).
	2	Bayer ADVIA 120 (Bayer Diagnostics, Tarrytown, NY).
Comment(s)	1,2	Data from outpatients. A computerized approach adapted from the Hoffmann technique was used to remove outliers and obtain the 2.5th and 97.5th percentiles.

CELLULAR HEMOGLOBIN CONCENTRATION DISTRIBUTION WIDTH (HDW)

	Male		Female	
Age	**n**	**g/dL**	**n**	**g/dL**
1–14 d	48	2.6–3.5	35	2.6–3.2
15–30 d	41	2.7–3.3	34	2.5–3.0
31–60 d	171	2.6–3.7	130	2.5–3.5
61–180 d	524	2.4–3.3	387	2.3–3.2
0.5–<2 y	2039	2.4–3.3	1688	2.3–3.2
2–<6 y	2382	2.4–3.2	2020	2.3–3.2
6–<12 y	1915	2.4–3.4	1786	2.3–3.3
12–<18 y	1738	2.3–3.3	2402	2.2–3.2
>18 y	1037	2.3–3.6	2477	2.2–3.2

Specimen Type(s)	Whole blood (K_2/EDTA anticoagulant)
Reference(s)	Brugnara C. Unpublished data.
Method(s)	Bayer ADVIA 120 (Bayer Diagnostics, Tarrytown, NY).
Comment(s)	Data from outpatients. A computerized approach adapted from the Hoffmann technique was used to remove outliers and obtain the 2.5th and 97.5th percentiles.

EOSINOPHIL COUNT (ABSOLUTE)

		Male			Female		
Test	Age	n	× 10^3/µL	× 10^9/L	n	× 10^3/µL	× 10^9/L
1	1–3 d	41	0–0.6	0–0.6	43	0–0.6	0–0.6
	4–7 d	26	0–0.7	0–0.7	25	0–0.6	0–0.6
	8–14 d	39	0–0.8	0–0.8	52	0–0.6	0–0.6
	15–30 d	125	0–0.8	0–0.8	85	0–0.7	0–0.7
	31–60 d	381	0–0.6	0–0.6	284	0–0.5	0–0.5
	61–180 d	1214	0–0.5	0–0.5	953	0–0.4	0–0.4
	0.5–<2 y	4935	0–0.3	0–0.3	3820	0–0.3	0–0.3
	2–<6 y	5641	0–0.3	0–0.3	4856	0–0.3	0–0.3
	6–<12 y	4690	0–0.4	0–0.4	4146	0–0.3	0–0.3
	12–<18 y	3515	0–0.4	0–0.4	4530	0–0.3	0–0.3
	≥18 y	1762	0–0.3	0–0.3	2744	0–0.3	0–0.3
2	1–3 d	172	0–0.3	0–0.3	114	0–0.2	0–0.2
	4–7 d	152	0–0.8	0–0.8	108	0–0.5	0–0.5
	8–14 d	180	0–0.6	0–0.6	124	0–0.6	0–0.6
	15–30 d	390	0–0.9	0–0.9	291	0–0.5	0–0.5
	31–60 d	74	0–0.5	0–0.5	184	0–0.2	0–0.2
	61–180 d	75	0–0.4	0–0.4	102	0–0.1	0–0.1
	0.5–<2 y	177	0–0.3	0–0.3	119	0–0.2	0–0.2
	2–<6 y	189	0–0.2	0–0.2	134	0–0.2	0–0.2
	6–<12 y	126	0–0.3	0–0.3	136	0–0.2	0–0.2
	12–18 y	140	0–0.3	0–0.3	154	0–0.2	0–0.2
3	1–14 d	38	0–0.7	0–0.7	28	0–0.6	0–0.6
	15–30 d	35	0–0.9	0–0.9	27	0–0.4	0–0.4
	31–60 d	146	0–0.6	0–0.6	109	0–0.5	0–0.5
	61–180 d	471	0–0.4	0–0.4	359	0–0.4	0–0.4
	0.5–<2 y	1558	0–0.4	0–0.4	1257	0–0.3	0–0.3
	2–<6 y	1646	0–0.4	0–0.4	1331	0–0.3	0–0.3
	6–<12 y	1600	0–0.5	0–0.5	1411	0–0.3	0–0.3
	12–<18 y	1430	0–0.3	0–0.3	1799	0–0.3	0–0.3
	>18 y	717	0–0.4	0–0.4	1092	0–0.3	0–0.3

Specimen Type(s)	1,2	Whole blood (K_3/EDTA anticoagulant)
	3	Whole blood (K_2/EDTA anticoagulant)
Reference(s)	1,2	Soldin SJ, Brugnara C, et al., eds. Pediatric reference ranges, 2nd ed. Washington, DC: AACC Press, 1997.
	3	Brugnara C. Unpublished data.
Method(s)	1	Bayer H*3 (Bayer Diagnostics, Tarrytown, NY).
	2	Coulter STKS (Beckman Coulter Inc., Brea, CA).
	3	Bayer ADVIA 120 (Bayer Diagnostics, Tarrytown, NY).
Comment(s)	1,3	Data from outpatients. A computerized approach adapted from the Hoffmann technique was used to remove outliers and obtain the 2.5th and 97.5th percentiles.
	2	Data from hospitalized patients (excluding hematology, oncology, and intensive care unit patients). A computerized approach adapted from the Hoffmann technique was used to remove outliers and obtain the 2.5th and 97.5th percentiles.

EOSINOPHIL COUNT (RELATIVE)

		Male			Female		
Test	**Age**	**n**	**%**	**Number Fraction**	**n**	**%**	**Number Fraction**
1	1–3 d	55	0–8.0	0–.080	54	0–5.6	0–.056
	4–7 d	35	0–5.9	0–.059	31	0–5.1	0–.051
	8–14 d	44	0–6.7	0–.067	61	0–5.6	0–.056
	15–30 d	148	0–6.4	0–.064	104	0–5.1	0–.051
	31–60 d	487	0–5.7	0–.057	374	0–4.7	0–.047
	61–180 d	1526	0–4.6	0–.046	1200	0–3.7	0–.037
	0.5–<2 y	6136	0–3.3	0–.033	4741	0–2.7	0–.027
	2–<6 y	6713	0–4.3	0–.043	5811	0–3.4	0–.034
	6–<12 y	5340	0–5.8	0–.058	4812	0–4.7	0–.047
	12–<18 y	4009	0–5.5	0–.055	5199	0–3.9	0–.039
	>18 y	2014	0–5.3	0–.053	3109	0–3.5	0–.035
2	1–3 d	172	0–2.9	0–.029	114	0–2.9	0–.029
	4–7 d	152	0–5.4	0–.054	108	0–7.5	0–.075
	8–14 d	180	0–5.0	0–.050	124	0–4.1	0–.041
	15–30 d	390	0–7.0	0–.070	290	0–5.4	0–.054
	31–60 d	74	0–5.4	0–.054	184	0–3.1	0–.031
	61–180 d	75	0–3.5	0–.035	102	0–2.8	0–.028
	0.5–<2 y	177	0–2.0	0–.020	119	0–3.1	0–.031
	2–<6 y	189	0–2.4	0–.024	134	0–3.3	0–.033
	6–<12 y	126	0–4.8	0–.048	136	0–4.0	0–.040
	12–18 y	140	0–2.8	0–.028	154	0–4.1	0–.041
3	1–14 d	38	0–6.1	0–.061	28	0–4.2	0–.042
	15–30 d	33	0–7.0	0–.070	27	0–4.0	0–.040
	31–60 d	148	0–5.7	0–.057	109	0–5.0	0–.050
	61–180 d	469	0–4.3	0–.043	357	0–3.8	0–.038
	0.5–<2 y	1555	0–3.5	0–.035	1254	0–2.8	0–.028
	2–<6 y	1636	0–4.2	0–.042	1317	0–3.9	0–.039
	6–<12 y	1487	0–5.6	0–.056	1395	0–4.2	0–.042
	12–<18 y	1413	0–4.5	0–.045	1770	0–3.7	0–.037
	>18 y	713	0–4.9	0–.049	1079	0–3.9	0–.039

Specimen Type(s)	1,2	Whole blood (K_3/EDTA anticoagulant)
	3	Whole blood (K_2/EDTA anticoagulant)
Reference(s)	1,2	Soldin SJ, Brugnara C, et al., eds. Pediatric reference ranges, 2nd ed. Washington, DC: AACC Press, 1997.
	3	Brugnara C. Unpublished data.
Method(s)	1	Bayer H*3 (Bayer Diagnostics, Tarrytown, NY).
	2	Coulter STKS (Beckman Coulter Inc., Brea, CA).
	3	Bayer ADVIA 120 (Bayer Diagnostics, Tarrytown, NY).
Comment(s)	1,3	Data from outpatients. A computerized approach adapted from the Hoffmann technique was used to remove outliers and obtain the 2.5th and 97.5th percentiles.
	2	Data from hospitalized patients (excluding hematology, oncology, and intensive care unit patients). A computerized approach adapted from the Hoffmann technique was used to remove outliers and obtain the 2.5th and 97.5th percentiles.

HEMATOCRIT

		Male			Female		
Test	Age	n	%	Volume Fraction	n	%	Volume Fraction
1	1–3 d	76	43.4–56.1	.434–.561	76	37.4–55.9	.374–.559
	4–7 d	49	40.2–54.7	.402–.547	39	39.1–56.7	.391–.567
	8–14 d	61	33.7–51.1	.337–.511	69	36.4–51.2	.364–.512
	15–30 d	168	29.7–44.2	.297–.442	116	30.6–44.7	.306–.447
	31–60 d	562	26.2–35.3	.262–.353	419	26.3–36.6	.263–.366
	61–180 d	1842	28.7–36.1	.287–.361	1364	28.5–36.1	.285–.361
	0.5–<2 y	8859	30.9–37.0	.309–.370	6993	31.2–37.2	.312–.372
	2–<6 y	11146	31.7–37.7	.317–.377	9750	32.0–37.1	.320–.371
	6–<12 y	7468	32.7–39.3	.327–.393	6799	33.0–39.6	.330–.396
	12–<18 y	5551	34.8–43.9	.348–.439	7995	34.0–40.7	.340–.407
	>18 y	3175	33.4–46.2	.334–.462	8608	33.0–41.0	.330–.410
2	1–3 d	216	33.5–52.7	.335–.527	143	38.4–56.0	.384–.560
	4–7 d	211	36.9–50.1	.369–.501	132	34.4–50.8	.344–.508
	8–14 d	226	33.4–49.0	.334–.490	151	32.9–47.5	.329–.475
	15–30 d	445	30.7–44.9	.307–.449	338	31.9–45.2	.319–.452
	31–60 d	81	27.0–35.8	.270–.358	197	27.1–41.9	.271–.419
	61–180 d	86	29.9–39.0	.299–.390	109	28.6–42.3	.286–.423
	0.5–<2 y	198	28.3–40.0	.283–.400	139	28.1–39.1	.281–.391
	2–<6 y	204	28.5[a]–37.9	.285[a]–.379	147	32.0[a]–39.8	.320[a]–.398
	6–<12 y	141	31.6[a]–39.5	.316[a]–.395	148	32.0[a]–40.0	.320[a]–.400
	12–18 y	155	27.3[a]–43.6	.273[a]–.436	172	27.9[a]–39.6	.279[a]–.396
3	1–14 d	48	36.2–58.5	.362–.585	35	39.1–58.5	.391–.585
	15–30 d	41	26.7–50.3	.267–.503	34	31.8–46.9	.318–.469
	31–60 d	172	25.2–37.1	.252–.371	130	27.2–41.6	.272–.416
	61–180 d	524	28.2–39.7	.282–.397	389	28.8–39.5	.288–.395
	0.5–<2 y	2041	30.8–39.1	.308–.391	1689	30.7–39.3	.307–.393
	2–<6 y	2386	31.5–40.4	.315–.404	2022	31.6–40.4	.316–.404
	6–<12 y	1919	32.0–41.9	.320–.419	1786	32.0–41.8	.320–.418
	12–<18 y	1738	32.3–46.8	.323–.468	2408	33.1–43.4	.331–.434
	>18 y	1038	30.0–48.9	.300–.489	2479	32.6–43.7	.326–.437

4	0–14 d	68	39.8–53.6	.398–.536	51	39.6–57.2	.396–.572
	15–30 d	58	30.5–45.0	.305–.450	42	32.0–44.5	.320–.445
	31–60 d	116	26.8–37.5	.268–.375	77	27.7–35.1	.277–.351
	61–180 d	305	28.6–37.2	.286–.372	231	29.5–37.1	.295–.371
	0.5–<2 y	1538	30.8–37.8	.308–.378	1343	30.9–37.9	.309–.379
	2–<6 y	1765	31.0–37.7	.310–.377	1680	31.2–37.8	.312–.378
	6–<12 y	1520	32.2–39.8	.322–.398	1304	32.4–39.5	.324–.395
	12–<18 y	1598	33.9–43.5	.339–.435	1944	33.4–40.4	.334–.404
	>18 y	275	36.2–46.3	.362–.463	486	32.9–41.2	.329–.412

Specimen Type(s)	1,2,4	Whole blood (K_3/EDTA anticoagulant)
	3	Whole blood (K_2/EDTA anticoagulant)
Reference(s)	1,2	Soldin SJ, Brugnara C, et al., eds. Pediatric reference ranges, 2nd ed. Washington, DC: AACC Press, 1997.
	3	Brugnara C. Unpublished data.
	4	Children's National Medical Center, Washington, DC. Unpublished data.
Method(s)	1	Bayer H*3 (Bayer Diagnostics, Tarrytown, NY).
	2	Coulter STKS (Beckman Coulter Inc., Brea, CA).
	3	Bayer ADVIA 120 (Bayer Diagnostics, Tarrytown, NY).
	4	Sysmex XE-2100 (Sysmex Corporation, Kobe, Japan).
Comment(s)	1,3	Data from outpatients. A computerized approach adapted from the Hoffmann technique was used to remove outliers and obtain the 2.5th and 97.5th percentiles (1). The upper 90% of the data was used for the Hoffman Plot for set 3.
	2	Data from hospitalized patients (excluding hematology, oncology, and intensive care unit patients). A computerized approach adapted from the Hoffmann technique was used to remove outliers and obtain the 2.5th and 97.5th percentiles. [a]To obtain the 2.5th percentile, the statistical technique was modified to use the central 60% of the data for the Hoffmann plot.
	4	Performed on the XE-2100 on outpatients and ER patients. Hematology oncology clinic patients were excluded. Data were analyzed using SAS statistical software. Outliers were removed according to Chauvenet's Criterion (1). This criterion uses an exclusion factor based on the number of values in the sample set and is reapplied until no more outliers are excluded. The remaining data was truncated based on a method described by Hoffmann—the top and bottom 10% of data, based on an assumed normal distribution, was discarded (2). The method was modified to allow for computerized handling of the data. The 2.5th and 97.5th percentiles were calculated on the remaining 80% of the data. 1. Young HD. Statistical treatment of experimental data. New York: McGraw-Hill, 1962:76–9, 162. 2. Hoffmann RG. Statistics in the practice of medicine. JAMA 1963;185:864–73.

HEMOGLOBIN

		Male			Female		
Test	Age	n	g/dL	g/L	n	g/dL	g/L
1	1–3 d	76	14.7–18.6	147–186	76	12.7–18.3	127–183
	4–7 d	49	13.4–17.9	134–179	38	12.2–18.7	122–187
	8–14 d	61	11.1–16.7	111–167	69	11.9–16.9	119–169
	15–30 d	168	9.9–14.9	99–149	116	10.5–14.7	105–147
	31–60 d	562	8.9–11.9	89–119	419	8.9–12.3	89–123
	61–180 d	1839	9.7–12.2	97–122	1361	9.7–12.0	97–120
	0.5–<2 y	8829	10.3–12.4	103–124	6917	10.4–12.4	104–124
	2–<6 y	11123	10.5–12.7	105–127	9729	10.7–12.7	107–127
	6–<12 y	7458	11.0–13.3	110–133	6778	10.9–13.3	109–133
	12–<18 y	5546	11.5–14.8	115–148	7989	11.2–13.6	112–136
	>18 y	3167	10.9–15.7	109–157	8600	10.7–13.5	107–135
2	1–3 d	217	12.2–17.9	122–179	144	13.2–18.4	132–184
	4–7 d	209	12.8–16.9	128–169	132	11.8–17.7	118–177
	8–14 d	224	11.8–16.8	118–168	151	11.5–16.3	115–163
	15–30 d	447	10.6–15.4	106–154	339	10.9–15.3	109–153
	31–60 d	82	9.0–12.1	90–121	197	9.2–14.4	92–144
	61–180 d	98	10.0–13.2	100–132	123	9.8–13.7	98–137
	0.5–<2 y	199	9.8–13.4	98–134	143	9.6–13.1	96–131
	2–<6 y	200	9.6[a]–12.8	96[a]–128	144	10.9[a]–13.4	109[a]–134
	6–<12 y	135	10.7[a]–13.5	107[a]–135	143	10.9[a]–13.7	109[a]–137
	12–18 y	147	9.5[a]–14.8	95[a]–148	174	9.5[a]–13.3	95[a]–133
3	1–14 d	48	12.2–19.9	122–199	35	13.6–18.8	136–188
	15–30 d	41	9.1–16.9	91–169	34	10.5–15.6	105–156
	31–60 d	172	8.7–12.7	87–127	130	9.4–13.5	94–135
	61–180 d	524	9.7–13.3	97–133	388	9.9–13.1	99–131
	0.5–<2 y	2037	10.3–13.1	103–131	1691	10.4–13.2	104–132
	2–<6 y	2384	10.7–13.8	107–138	2021	10.6–13.8	106–138
	6–<12 y	1909	10.8–14.4	108–144	1785	10.6–14.4	106–144
	12–<18 y	1740	10.7–16.0	107–160	2405	10.9–14.6	109–146
	>18 y	1037	9.9–16.5	99–165	2477	10.6–14.5	106–145

4	1–14 d	60	13.9–19.1	139–191	48	13.4–20.0	134–200
	15–30 d	55	10.0–15.3	100–153	37	10.8–14.6	108–146
	31–60 d	110	8.9–12.7	89–127	71	9.2–11.4	92–114
	61–180 d	277	9.6–12.4	96–124	209	9.9–12.4	99–124
	0.5–<2 y	1350	10.1–12.5	101–125	1171	10.2–12.7	102–127
	2–<6 y	1383	10.2–12.7	102–127	1281	10.2–12.7	102–127
	6–<12 y	1420	10.7–13.4	107–134	1207	10.6–13.2	106–132
	12–<18 y	1569	11.0–14.5	110–145	1872	10.8–13.3	108–133
	>18 y	268	11.9–15.4	119–154	471	10.6–13.5	106–135

Specimen Type(s)	1,2,4	Whole blood (K_3/EDTA anticoagulant)
	3	Whole blood (K_2/EDTA anticoagulant)
Reference(s)	1,2	Soldin SJ, Brugnara C, et al., eds. Pediatric reference ranges, 2nd ed. Washington, DC: AACC Press, 1997.
	3	Brugnara C. Unpublished data.
	4	Children's National Medical Center, Washington, DC. Unpublished data.
Method(s)	1	Bayer H*3 (Bayer Diagnostics, Tarrytown, NY).
	2	Coulter STKS (Beckman Coulter Inc., Brea, CA).
	3	Bayer ADVIA 120 (Bayer Diagnostics, Tarrytown, NY).
	4	Sysmex XE-2100 (Sysmex Corporation, Kobe, Japan).
Comment(s)	1,3	Data from outpatients. A computerized approach adapted from the Hoffmann technique was used to remove outliers and obtain the 2.5th and 97.5th percentiles (1). The upper 90% of the data was used for the Hoffman Plot for data set 3.
	2	Data from hospitalized patients (excluding hematology, oncology, and intensive care unit patients). A computerized approach adapted from the Hoffmann technique was used to remove outliers and obtain the 2.5th and 97.5th percentiles.
		[a]To obtain the 2.5th percentile, the statistical technique was modified to use the central 60% of the data for the Hoffmann plot.
	4	Performed on the XE-2100 on outpatients and ER patients. Hematology oncology clinic patients were excluded. Data were analyzed using SAS statistical software. Outliers were removed according to Chauvenet's Criterion (1). This criterion uses an exclusion factor based on the number of values in the sample set and is reapplied until no more outliers are excluded. The remaining data was truncated based on a method described by Hoffmann—the top and bottom 10% of data, based on an assumed normal distribution, was discarded (2). The method was modified to allow for computerized handling of the data. The 2.5th and 97.5th percentiles were calculated on the remaining 80% of the data.
		1. Young HD. Statistical treatment of experimental data. New York: McGraw-Hill, 1962:76–9, 162.
		2. Hoffmann RG. Statistics in the practice of medicine. JAMA 1963;185:864–73.

IMMATURE RETICULOCYTE FRACTION (IRF)

	Male			Female		
Age	n	Percent	Fraction	n	Percent	Fraction
0–6 mo	30	11.0–26.9	.110–.269	15	10.7–30.5	0.107–0.305
0.5–<2 y	57	8.5–30.6	.085–.306	61	8.4–27.3	0.084–0.273
2–<6 y	86	7.1–33.0	.071–.330	63	5.9–26.6	0.059–0.266
6–<12 y	78	7.3–30.3	.073–.303	61	8.3–30.0	0.083–0.300
12–<18 y	88	5.2–29.4	.052–.294	70	7.5–27.3	0.075–0.273

Specimen Type(s)	Whole blood (K_3/EDTA anticoagulant)
Reference(s)	Children's National Medical Center, Washington, DC. Unpublished data.
Method(s)	Sysmex XE-2100 (Sysmex Corporation, Kobe, Japan).
Comment(s)	Performed on the XE-2100 on outpatients and ER patients. Hematology oncology clinic patients were excluded. Data were analyzed using SAS statistical software. Outliers were removed according to Chauvenet's Criterion (1). This criterion uses an exclusion factor based on the number of values in the sample set and is reapplied until no more outliers are excluded. The remaining data was truncated based on a method described by Hoffmann—the top and bottom 10% of data, based on an assumed normal distribution, was discarded (2). The method was modified to allow for computerized handling of the data. The 2.5th and 97.5th percentiles were calculated on the remaining 80% of the data. 1. Young HD. Statistical treatment of experimental data. New York: McGraw-Hill, 1962:76–9, 162. 2. Hoffmann RG. Statistics in the practice of medicine. JAMA 1963;185:864–73.

LYMPHOCYTE COUNT (ABSOLUTE)

		Male			Female		
Test	**Age**	**n**	**× 10^3/µL**	**× 10^9/L**	**n**	**× 10^3/µL**	**× 10^9/L**
1	1–3 d	47	2.2–5.4	2.2–5.4	50	2.8–5.3	2.8–5.3
	4–7 d	32	4.3–7.7	4.3–7.7	29	4.9–7.0	4.9–7.0
	8–14 d	39	4.2–7.4	4.2–7.4	58	4.4–8.3	4.4–8.3
	15–30 d	137	3.9–8.5	3.9–8.5	98	4.1–8.9	4.1–8.9
	31–60 d	463	3.3–8.3	3.3–8.3	358	3.2–9.1	3.2–9.1
	61–180 d	1475	2.8–8.3	2.8–8.3	1166	2.8–8.4	2.8–8.4
	0.5–<2 y	5941	1.9–6.8	1.9–6.8	4608	1.2–7.0	1.2–7.0
	2–<6 y	6570	1.3–4.7	1.3–4.7	5672	1.4–4.7	1.4–4.7
	6–<12 y	5343	1.1–3.4	1.1–3.4	4719	1.1–3.5	1.1–3.5
	12–<18 y	3953	1.0–2.8	1.0–2.8	5109	1.1–2.8	1.1–2.8
	>18 y	1969	0.9–2.5	0.9–2.5	3054	1.1–2.7	1.1–2.7
2	1–3 d	173	0.9–5.5	0.9–5.5	114	1.2–5.1	1.2–5.1
	4–7 d	152	1.1–7.8	1.1–7.8	108	1.3–5.4	1.3–5.4
	8–14 d	180	1.3–7.5	1.3–7.5	124	0.8–8.5	0.8–8.5
	15–30 d	390	0.8–8.3	0.8–8.3	291	0.9–8.0	0.9–8.0
	31–60 d	74	1.7–8.4	1.7–8.4	184	1.2–8.0	1.2–8.0
	61–180 d	75	1.3–8.8	1.3–8.8	102	2.1–7.4	2.1–7.4
	0.5–<2 y	177	1.1–6.5	1.1–6.5	119	1.6–7.5	1.6–7.5
	2–<6 y	189	0.5–4.4	0.5–4.4	134	1.2–5.0	1.2–5.0
	6–<12 y	126	0.8–3.0	0.8–3.0	136	0.9–3.4	0.9–3.4
	12–18 y	140	0.4–2.5	0.4–2.5	154	0.8–3.2	0.8–3.2
3	1–14 d	38	3.9–7.2	3.9–7.2	28	3.0–6.5	3.0–6.5
	15–30 d	33	3.1–7.0	3.1–7.0	27	3.2–7.8	3.2–7.8
	31–60 d	148	2.6–6.8	2.6–6.8	109	2.2–7.1	2.2–7.1
	61–180 d	469	2.5–7.8	2.5–7.8	357	2.3–7.7	2.3–7.7
	0.5–<2 y	1557	1.7–6.5	1.7–6.5	1257	1.7–6.7	1.7–6.7
	2–<6 y	1637	1.3–4.6	1.3–4.6	1315	1.2–4.5	1.2–4.5
	6–<12 y	1490	1.0–3.4	1.0–3.4	1391	1.2–3.4	1.2–3.4
	12–<18 y	1411	1.0–2.8	1.0–2.8	1774	1.1–2.8	1.1–2.8
	>18 y	713	0.8–2.5	0.8–2.5	1077	0.9–2.6	0.9–2.6

Specimen Type(s)	1,2	Whole blood (K_3/EDTA anticoagulant)
	3	Whole blood (K_2/EDTA anticoagulant)
Reference(s)	1,2	Soldin SJ, Brugnara C, et al., eds. Pediatric reference ranges, 2nd ed. Washington, DC: AACC Press, 1997.
	3	Brugnara C. Unpublished data.
Method(s)	1	Bayer H*3 (Bayer Diagnostics, Tarrytown, NY).
	2	Coulter STKS (Beckman Coulter Inc., Brea, CA).
	3	Bayer ADVIA 120 (Bayer Diagnostics, Tarrytown, NY).
Comment(s)	1,3	Data from outpatients. A computerized approach adapted from the Hoffmann technique was used to remove outliers and obtain the 2.5th and 97.5th percentiles.
	2	Data from hospitalized patients (excluding hematology, oncology, and intensive care unit patients). A computerized approach adapted from the Hoffmann technique was used to remove outliers and obtain the 2.5th and 97.5th percentiles.

LYMPHOCYTE COUNT (RELATIVE)

		Male			Female		
Test	Age	n	%	Number Fraction	n	%	Number Fraction
1	1–3 d	48	25.9–56.5	.259–.565	50	28.4–54.6	.284–.546
	4–7 d	33	39.3–60.7	.393–.607	29	38.8–64.1	.388–.641
	8–14 d	43	40.2–62.2	.402–.622	58	44.6–67.3	.446–.673
	15–30 d	137	41.3–65.4	.413–.654	98	35.1–67.4	.351–.674
	31–60 d	463	39.5–69.7	.395–.697	359	36.7–69.8	.367–.698
	61–180 d	1475	32.0–68.5	.320–.685	1165	30.4–68.9	.304–.689
	0.5–<2 y	5945	19.8–63.7	.198–.637	4614	19.8–62.8	.198–.628
	2–<6 y	6581	14.1–55.0	.141–.550	5680	15.6–55.6	.156–.556
	6–<12 y	5355	14.3–47.9	.143–.479	4730	13.1–48.4	.131–.484
	12–<18 y	3962	13.4–42.8	.134–.428	5123	14.1–41.3	.141–.413
	>18 y	1970	12.8–43.9	.128–.439	3062	15.4–39.5	.154–.395
2	1–3 d	173	16.8–46.4	.168–.464	114	13.4–40.7	.134–.407
	4–7 d	152	18.8–51.4	.188–.514	108	18.4–52.9	.184–.529
	8–14 d	180	13.0–43.6	.130–.436	124	19.0–54.7	.190–.547
	15–30 d	390	10.4–51.9	.104–.519	292	17.0–58.7	.170–.587
	31–60 d	302	22.7–53.9	.227–.539	184	21.3–63.6	.213–.636
	61–180 d	25	19.9–53.8	.199–.538	41	25.7–52.9	.257–.529
	0.5–<2 y	91	18.6–58.1	.186–.581	64	16.4–56.3	.164–.563
	2–<6 y	189	11.0–50.1	.110–.501	134	14.3–50.6	.143–.506
	6–<12 y	126	14.1–43.1	.141–.431	136	13.1–45.2	.131–.452
	12–18 y	140	6.5–32.9	.065–.329	154	12.9–40.6	.129–.406
3	1–14 d	38	26.2–52.8	.262–.528	28	25.0–51.9	.250–.519
	15–30 d	33	39.8–60.6	.398–.606	27	34.8–66.5	.348–.665
	31–60 d	148	34.0–67.3	.340–.673	109	30.4–64.7	.304–.647
	61–180 d	469	28.0–64.6	.280–.646	357	24.0–64.0	.240–.640
	0.5–<2 y	1557	17.0–61.1	.170–.611	1258	18.4–61.0	.184–.610
	2–<6 y	1637	13.5–52.8	.135–.528	1317	12.5–51.4	.125–.514
	6–<12 y	1491	12.2–44.9	.122–.449	1395	13.7–43.6	.137–.436
	12–<18 y	1415	13.0–41.1	.130–.411	1770	14.5–39.6	.145–.396
	>18 y	713	11.5–36.3	.115–.363	1079	12.1–35.8	.121–.358

Specimen Type(s)	1,2	Whole blood (K_3/EDTA anticoagulant)
	3	Whole blood (K_2/EDTA anticoagulant)
Reference(s)	1,2	Soldin SJ, Brugnara C, et al., eds. Pediatric reference ranges, 2nd ed. Washington, DC: AACC Press, 1997.
	3	Brugnara C. Unpublished data.
Method(s)	1	Bayer H*3 (Bayer Diagnostics, Tarrytown, NY).
	2	Coulter STKS (Beckman Coulter Inc., Brea, CA).
	3	Bayer ADVIA 120 (Bayer Diagnostics, Tarrytown, NY).
Comment(s)	1,3	Data from outpatients. A computerized approach adapted from the Hoffmann technique was used to remove outliers and obtain the 2.5th and 97.5th percentiles.
	2	Data from hospitalized patients (excluding hematology, oncology, and intensive care unit patients). A computerized approach adapted from the Hoffmann technique was used to remove outliers and obtain the 2.5th and 97.5th percentiles. To obtain the 2.5th and 97.5th percentiles, the statistical technique was modified to use the central 60% of the data for the Hoffmann plot.

MEAN CORPUSCULAR HEMOGLOBIN (MCH)

		Male		Female	
Test	**Age**	**n**	**pg**	**n**	**pg**
1	1–3 d	75	32.5–36.5	70	32.6–37.8
	4–7 d	49	31.1–36.5	38	32.2–36.6
	8–14 d	61	30.6–35.7	69	31.7–36.5
	15–30 d	168	30.1–33.8	116	30.8–34.6
	31–60 d	561	28.4–32.6	420	28.6–32.9
	61–180 d	1838	24.5–29.1	1362	24.7–29.6
	0.5–<2 y	8842	23.2–27.5	6926	23.5–27.6
	2–<6 y	11140	24.1–28.4	9746	24.3–28.6
	6–<12 y	7465	25.4–29.4	6806	25.4–29.6
	12–<18 y	5545	25.8–30.3	7993	26.1–30.4
	>18 y	3171	27.2–31.7	8604	26.2–31.0
2	1–3 d	174	31.3–38.6	115	31.4–39.4
	4–7 d	155	29.9–36.2	108	30.8–36.8
	8–14 d	182	29.1–35.7	124	30.3–35.0
	15–30 d	393	28.8–33.6	291	29.2–34.3
	31–60 d	74	28.0–30.5	185	29.0–32.7
	61–180 d	75	25.8–29.2	102	25.8–30.1
	0.5–<2 y	177	24.2–30.0	119	23.4–30.0
	2–<6 y	190	24.6–30.9	135	25.3–29.9
	6–<12 y	128	26.3–30.3	138	27.4–33.2
	12–18 y	140	25.9–31.0	154	26.7–31.7
3	1–14 d	48	32.2–35.4	35	32.4–36.5
	15–30 d	41	30.0–33.7	34	29.7–34.4
	31–60 d	172	28.6–31.8	130	28.6–32.2
	61–180 d	524	24.2–29.0	386	24.8–29.2
	0.5–<2 y	2039	23.2–27.3	1686	23.4–27.4
	2–<6 y	2382	24.2–28.1	2019	24.1–28.5
	6–<12 y	1912	25.6–29.4	1788	25.2–29.7
	2–<18 y	1740	25.7–30.2	2403	25.9–30.2
	>18 y	1038	27.1–31.8	2479	25.8–31.3

4	1–14 d	62	31.3–35.6	44	31.1–35.9
	15–30 d	58	29.9–34.1	39	30.4–35.3
	31–60 d	104	27.8–32.0	78	28.0–32.5
	61–180 d	292	24.4–28.9	219	24.4–29.5
	0.5–<2 y	1262	22.7–27.2	1086	23.2–27.5
	2–<6 y	1266	23.7–28.3	1132	23.7–28.6
	6–<12 y	1341	24.9–29.2	1146	24.8–29.5
	2–<18 y	1536	25.2–30.2	1848	24.8–30.2
	>18 y	268	26.5–31.4	475	25.3–30.9

Specimen Type(s)	1,2,4	Whole blood (K_3/EDTA anticoagulant)
	3	Whole blood (K_2/EDTA anticoagulant)
Reference(s)	1,2	Soldin SJ, Brugnara C, et al., eds. Pediatric reference ranges, 2nd ed. Washington, DC: AACC Press, 1997.
	3	Brugnara C. Unpublished data.
	4	Children's National Medical Center, Washington, DC. Unpublished data.
Method(s)	1	Bayer H*3 (Bayer Diagnostics, Tarrytown, NY).
	2	Coulter STKS (Beckman Coulter Inc., Brea, CA).
	3	Bayer ADVIA 120 (Bayer Diagnostics, Tarrytown, NY).
	4	Sysmex XE-2100 (Sysmex Corporation, Kobe, Japan).
Comment(s)	1,3	Data from outpatients. A computerized approach adapted from the Hoffmann technique was used to remove outliers and obtain the 2.5th and 97.5th percentiles.
	2	Data from hospitalized patients (excluding hematology, oncology, and intensive care unit patients). A computerized approach adapted from the Hoffmann technique was used to remove outliers and obtain the 2.5th and 97.5th percentiles.
	4	Performed on the XE-2100 on outpatients and ER patients. Hematology oncology clinic patients were excluded. Data were analyzed using SAS statistical software. Outliers were removed according to Chauvenet's Criterion (1). This criterion uses an exclusion factor based on the number of values in the sample set and is reapplied until no more outliers are excluded. The remaining data was truncated based on a method described by Hoffmann—the top and bottom 10% of data, based on an assumed normal distribution, was discarded (2). The method was modified to allow for computerized handling of the data. The 2.5th and 97.5th percentiles were calculated on the remaining 80% of the data. 1. Young HD. Statistical treatment of experimental data. New York: McGraw-Hill, 1962:76–9, 162. 2. Hoffmann RG. Statistics in the practice of medicine. JAMA 1963;185:864–73.

MEAN CORPUSCULAR HEMOGLOBIN CONCENTRATION (MCHC)

		Male			Female		
Test	**Age**	**n**	**%**	**Concentration Fraction**	**n**	**%**	**Concentration Fraction**
1	1–3 d	76	32.1–34.2	.321–.342	76	31.6–34.7	.316–.347
	4–7 d	49	32.0–34.6	.320–.346	39	31.8–34.6	.318–.346
	8–14 d	60	32.0–35.0	.320–.350	69	32.1–34.8	.321–.348
	15–30 d	167	32.3–35.0	.323–.350	116	32.0–35.0	.320–.350
	31–60 d	562	32.5–35.5	.325–.355	419	32.2–35.3	.322–.353
	61–180 d	1389	32.0–35.1	.320–.351	1363	32.0–35.1	.320–.351
	0.5–<2 y	8856	31.9–35.0	.319–.350	6931	31.8–34.8	.318–.348
	2–<6 y	11149	31.9–35.1	.319–.351	9744	31.9–35.0	.319–.350
	6–<12 y	7465	32.2–35.2	.322–.352	6803	31.9–35.0	.319–.350
	12–<18 y	5545	31.8–35.0	.318–.350	7991	31.6–34.7	.316–.347
	>18 y	3170	31.9–35.1	.319–.351	8607	31.4–34.3	.314–.343
2	1–3 d	174	33.5–34.8	.335–.348	115	33.3–34.5	.333–.345
	4–7 d	155	33.5–35.0	.335–.350	108	33.6–35.0	.336–.350
	8–14 d	182	33.5–35.0	.335–.350	124	33.5–35.2	.335–.352
	15–30 d	393	33.5–35.0	.335–.350	291	33.5–35.0	.335–.350
	31–60 d	74	33.5–34.6	.335–.346	185	33.5–34.8	.335–.348
	61–180 d	75	32.9–34.0	.329–.340	102	33.1–34.4	.331–.344
	0.5–<2 y	177	32.9–34.5	.329–.345	119	32.7–34.6	.327–.346
	2–<6 y	190	32.7–34.7	.327–.347	135	32.9–34.6	.329–.346
	6–<12 y	128	33.2–34.8	.332–.348	138	33.3–34.8	.333–.348
	12–18 y	140	33.4–34.6	.334–.346	154	33.2–34.7	.332–.347
3	1–14 d	48	32.8–35.3	.328–.353	35	32.9–34.4	.329–.344
	15–30 d	41	33.0–35.3	.330–.353	34	33.3–35.1	.333–.351
	31–60 d	172	33.0–35.8	.330–.358	130	33.4–35.5	.334–.355
	61–180 d	524	32.6–35.3	.326–.353	388	32.6–35.4	.326–.354
	0.5–<2 y	2041	32.4–35.0	.324–.350	1690	32.4–34.9	.324–.349
	2–<6 y	2386	32.8–35.3	.328–.353	2022	32.4–35.1	.324–.351
	6–<12 y	1917	32.8–35.3	.328–.353	1789	32.3–35.2	.323–.352
	12–<18 y	1740	32.3–35.2	.323–.352	2405	32.0–34.7	.320–.347
	>18 y	1038	32.0–34.9	.320–.349	2478	31.6–34.4	.316–.344

4	0–14 d	64	33.0–35.7	49	33.4–35.4
	15–30 d	59	32.7–35.1	39	33.2–35.0
	31–60 d	116	32.3–34.8	76	32.5–34.9
	61–180 d	301	31.9–34.4	227	32.1–34.4
	0.5–<2 y	1260	31.6–34.4	1088	31.9–34.2
	2–<6 y	1273	32.0–34.7	1143	31.8–34.6
	6–<12 y	1360	32.2–34.9	1148	31.8–34.6
	12–<18 y	1553	31.8–34.8	1845	31.5–34.2
	>18 y	271	31.9–34.8	483	31.0–34.1

Specimen Type(s)	1,2,4	Whole blood (K_3/EDTA anticoagulant)
	3	Whole blood (K_2/EDTA anticoagulant)
Reference(s)	1,2	Soldin SJ, Brugnara C, et al., eds. Pediatric reference ranges, 2nd ed. Washington, DC: AACC Press, 1997.
	3	Brugnara C. Unpublished data.
	4	Children's National Medical Center, Washington, DC. Unpublished data.
Method(s)	1	Bayer H*3 (Bayer Diagnostics, Tarrytown, NY).
	2	Coulter STKS (Beckman Coulter Inc., Brea, CA).
	3	Bayer ADVIA 120 (Bayer Diagnostics, Tarrytown, NY).
	4	Sysmex XE-2100 (Sysmex Corporation, Kobe, Japan).
Comment(s)	1,3	Data from outpatients. A computerized approach adapted from the Hoffmann technique was used to remove outliers and obtain the 2.5th and 97.5th percentiles.
	2	Data from hospitalized patients (excluding hematology, oncology, and intensive care unit patients). A computerized approach adapted from the Hoffmann technique was used to remove outliers and obtain the 2.5th and 97.5th percentiles.
	4	Performed on the XE-2100 on outpatients and ER patients. Hematology oncology clinic patients were excluded. Data were analyzed using SAS statistical software. Outliers were removed according to Chauvenet's Criterion (1). This criterion uses an exclusion factor based on the number of values in the sample set and is reapplied until no more outliers are excluded. The remaining data was truncated based on a method described by Hoffmann—the top and bottom 10% of data, based on an assumed normal distribution, was discarded (2). The method was modified to allow for computerized handling of the data. The 2.5th and 97.5th percentiles were calculated on the remaining 80% of the data. 1. Young HD. Statistical treatment of experimental data. New York: McGraw-Hill, 1962:76–9, 162. 2. Hoffmann RG. Statistics in the practice of medicine. JAMA 1963;185:864–73.

MEAN CORPUSCULAR VOLUME (MCV)

		Male			Female		
Test	**Age**	**n**	**μm^3**	**fL**	**n**	**μm^3**	**fL**
1	1–3 d	76	97.3–109.8	97.3–109.8	76	99.4–113.8	99.4–113.8
	4–7 d	49	95.5–109.3	95.5–109.3	39	97.9–111.6	97.9–111.6
	8–14 d	61	93.1–105.4	93.1–105.4	69	97.3–109.3	97.3–109.3
	15–30 d	168	88.7–101.2	88.7–101.2	116	91.8–102.5	91.8–102.5
	31–60 d	562	84.6–95.4	84.6–95.4	418	85.0–96.9	85.0–96.9
	61–180 d	1841	73.6–86.6	73.6–86.6	1363	74.7–87.6	74.7–87.6
	0.5–<2 y	8852	70.5–81.2	70.5–81.2	6927	71.5–81.8	71.5–81.8
	2–<6 y	11149	72.7–83.6	72.7–83.6	9750	73.8–84.3	73.8–84.3
	6–<12 y	7467	75.9–86.5	75.9–86.5	6795	76.8–87.6	76.8–87.6
	12–<18 y	5551	77.9–89.8	77.9–89.8	7995	79.4–91.0	79.4–91.0
	>18 y	3173	81.3–94.2	81.3–94.2	8608	80.8–93.4	80.8–93.4
2	1–3 d	174	93.0–113.4	93.0–113.4	117	92.4–115.4	92.4–115.4
	4–7 d	155	88.1–106.5	88.1–106.5	108	90.6–108.3	90.6–108.3
	8–14 d	182	85.1–104.5	85.1–104.5	124	87.9–101.8	87.9–101.8
	15–30 d	393	83.4–97.6	83.4–97.6	291	84.5–99.7	84.5–99.7
	31–60 d	74	82.0–89.8	82.0–89.8	185	85.5–95.3	85.5–95.3
	61–180 d	75	76.5–87.8	76.5–87.8	102	77.5–89.4	77.5–89.4
	0.5–<2 y	177	72.3–88.7	72.3–88.7	119	70.6–87.9	70.6–87.9
	2–<6 y	190	73.6–90.0	73.6–90.0	135	75.6–88.0	75.6–88.0
	6–<12 y	128	77.3–88.3	77.3–88.3	138	80.7–95.6	80.7–95.6
	12–18 y	140	76.3–90.1	76.3–90.1	154	79.1–91.7	79.1–91.7
3	1–14 d	48	95.9–100.9	95.9–100.9	35	98.0–104.2	98.0–104.2
	15–30 d	41	80.3–100.1	80.3–100.1	34	88.9–98.4	88.9–98.4
	31–60 d	172	83.9–91.8	83.9–91.8	130	82.9–93.8	82.9–93.8
	61–180 d	524	71.8–85.1	71.8–85.1	389	73.8–85.8	73.8–85.8
	0.5–<2 y	2041	69.5–79.7	69.5–79.7	1689	70.9–80.1	70.9–80.1
	2–<6 y	2385	72.4–82.2	72.4–82.2	2022	72.7–83.1	72.7–83.1
	6–<12 y	1917	75.1–85.9	75.1–85.9	1784	75.6–86.8	75.6–86.8
	12–<18 y	1740	76.9–88.8	76.9–88.8	2402	78.3–89.7	78.3–89.7
	>18 y	1038	80.6–95.0	80.6–95.0	2478	79.5–93.9	79.5–93.9

4	0–14 d	63	91.3–103.1	91.3–103.1	48	92.7–106.4	92.7–106.4
	15–30 d	56	89.4–99.7	89.4–99.7	39	90.1–103.0	90.1–103.0
	31–60 d	110	84.3–94.2	84.3–94.2	78	83.4–96.4	83.4–96.4
	61–180 d	298	74.1–87.5	74.1–87.5	229	74.8–88.3	74.8–88.3
	0.5–<2 y	1260	69.5–81.7	69.5–81.7	1081	71.3–82.6	71.3–82.6
	2–<6 y	1279	71.3–84.0	71.3–84.0	1133	72.3–85.0	72.3–85.0
	6–<12 y	1349	74.4–86.1	74.4–86.1	1140	75.9–87.6	75.9–87.6
	12–<18 y	1543	76.7–89.2	76.7–89.2	1840	76.9–90.6	76.9–90.6
	>18 y	273	80.0–93.6	80.0–93.6	477	77.7–93.7	77.7–93.7

Specimen Type(s)	1,2,4	Whole blood (K_3/EDTA anticoagulant)
	3	Whole blood (K_2/EDTA anticoagulant)
Reference(s)	1,2	Soldin SJ, Brugnara C, et al., eds. Pediatric reference ranges, 2nd ed. Washington, DC: AACC Press, 1997.
	3	Brugnara C. Unpublished data.
	4	Children's National Medical Center, Washington, DC. Unpublished data.
Method(s)	1	Bayer H*3 (Bayer Diagnostics, Tarrytown, NY).
	2	Coulter STKS (Beckman Coulter Inc., Brea, CA).
	3	Bayer ADVIA 120 (Bayer Diagnostics, Tarrytown, NY).
	4	Sysmex XE-2100 (Sysmex Corporation, Kobe, Japan).
Comment(s)	1,3	Data from outpatients. A computerized approach adapted from the Hoffmann technique was used to remove outliers and obtain the 2.5th and 97.5th percentiles.
	2	Data from hospitalized patients (excluding hematology, oncology, and intensive care unit patients). A computerized approach adapted from the Hoffmann technique was used to remove outliers and obtain the 2.5th and 97.5th percentiles.
	4	Performed on the XE-2100 on outpatients and ER patients. Hematology oncology clinic patients were excluded. Data were analyzed using SAS statistical software. Outliers were removed according to Chauvenet's Criterion (1). This criterion uses an exclusion factor based on the number of values in the sample set and is reapplied until no more outliers are excluded. The remaining data was truncated based on a method described by Hoffmann—the top and bottom 10% of data, based on an assumed normal distribution, was discarded (2). The method was modified to allow for computerized handling of the data. The 2.5th and 97.5th percentiles were calculated on the remaining 80% of the data. 1. Young HD. Statistical treatment of experimental data. New York: McGraw-Hill, 1962:76–9, 162. 2. Hoffmann RG. Statistics in the practice of medicine. JAMA 1963;185:864–73.

MEAN PLATELET VOLUME (MPV)

		Male			Female		
Test	Age	n	μm^3	fL	n	μm^3	fL
1	1–3 d	170	7.1–8.4	7.1–8.4	115	7.3–8.6	7.3–8.6
	4–7 d	151	7.5–9.2	7.5–9.2	107	7.8–9.3	7.8–9.3
	8–14 d	179	8.1–10.0	8.1–10.0	124	8.2–10.1	8.2–10.1
	15–30 d	387	8.0–10.6	8.0–10.6	287	8.0–10.9	8.0–10.9
	31–60 d	74	7.0–11.3	7.0–11.3	184	7.4–9.7	7.4–9.7
	61–180 d	75	6.8–9.0	6.8–9.0	102	7.2–8.9	7.2–8.9
	0.5–<2 y	177	7.1–9.3	7.1–9.3	119	7.0–9.3	7.0–9.3
	2–<6 y	190	7.1–9.3	7.1–9.3	135	6.9–8.8	6.9–8.8
	6–<12 y	127	7.2–9.4	7.2–9.4	135	7.1–9.2	7.1–9.2
	12–<18 y	140	7.3–9.7	7.3–9.7	154	7.5–9.3	7.5–9.3
2	1–14 d	46	7.6–9.9	7.6–9.9	35	7.5–9.3	7.5–9.3
	15–30 d	47	7.1–9.9	7.1–9.9	36	7.7–9.4	7.7–9.4
	31–60 d	167	7.4–9.1	7.4–9.1	129	7.5–9.5	7.5–9.5
	61–180 d	534	7.2–9.0	7.2–9.0	390	7.2–8.8	7.2–8.8
	0.5–<2 y	2056	7.0–8.8	7.0–8.8	1709	7.1–8.8	7.1–8.8
	2–<6 y	2423	7.1–8.9	7.1–8.9	2059	7.1–8.9	7.1–8.9
	6–<12 y	1933	7.3–9.1	7.3–9.1	1833	7.3–9.3	7.3–9.3
	12–<18 y	1802	7.5–9.6	7.5–9.6	2458	7.5–9.6	7.5–9.6
	>18 y	1039	7.6–9.6	7.6–9.6	2526	7.7–9.8	7.7–9.8
3	0–14 d	57	10.2–11.9	10.2–11.9	41	10.4–12.0	10.4–12.0
	15–30 d	51	10.1–12.1	10.1–12.1	38	10.0–12.2	10.0–12.2
	31–60 d	100	9.2–10.8	9.2–10.8	69	9.4–11.1	9.4–11.1
	61–180 d	264	8.9–10.6	8.9–10.6	204	9.0–10.9	9.0–10.9
	0.5–<2 y	1100	8.7–10.5	8.7–10.5	981	8.8–10.6	8.8–10.6
	2–<6 y	1138	9.0–10.9	9.0–10.9	999	8.9–11.0	8.9–11.0
	6–<12 y	1248	9.2–11.4	9.2–11.4	1057	9.3–11.3	9.3–11.3
	12–<18 y	1403	9.6–11.8	9.6–11.8	1705	9.6–11.7	9.6–11.7
	>18 y	240	9.7–11.9	9.7–11.9	431	9.6–12.0	9.6–12.0

Specimen Type(s)	1,3	Whole blood (K_3/EDTA anticoagulant)
	2	Whole blood (K_2/EDTA anticoagulant)
Reference(s)	1	Soldin SJ, Brugnara C, et al., eds. Pediatric reference ranges, 2nd ed. Washington, DC: AACC Press, 1997.
	2	Brugnara C. Unpublished data.
	3	Children's National Medical Center, Washington, DC. Unpublished data.
Method(s)	1	Coulter STKS (Beckman Coulter Inc., Brea, CA).
	2	Bayer ADVIA 120 (Bayer Diagnostics, Tarrytown, NY).
	3	Sysmex XE-2100 (Sysmex Corporation, Kobe, Japan).
Comment(s)	1	Data from hospitalized patients (excluding hematology, oncology, and intensive care unit patients). A computerized approach adapted from the Hoffmann technique was used to remove outliers and obtain the 2.5th and 97.5th percentiles.
	2	Data from outpatients. A computerized approach adapted from the Hoffmann technique was used to remove outliers and obtain the 2.5th and 97.5th percentiles.
	3	Performed on the XE-2100 on outpatients and ER patients. Hematology oncology clinic patients were excluded. Data were analyzed using SAS statistical software. Outliers were removed according to Chauvenet's Criterion (1). This criterion uses an exclusion factor based on the number of values in the sample set and is reapplied until no more outliers are excluded. The remaining data was truncated based on a method described by Hoffmann—the top and bottom 10% of data, based on an assumed normal distribution, was discarded (2). The method was modified to allow for computerized handling of the data. The 2.5th and 97.5th percentiles were calculated on the remaining 80% of the data. 1. Young HD. Statistical treatment of experimental data. New York: McGraw-Hill, 1962:76–9, 162. 2. Hoffmann RG. Statistics in the practice of medicine. JAMA 1963;185:864–73.

MONOCYTE COUNT (ABSOLUTE)

	Male			Female		
Age	n	× 10^3/µL	× 10^9/L	n	× 10^3/µL	× 10^9/L
1–3 d	172	0.2–1.8	0.2–1.8	114	0.2–2.2	0.2–2.2
4–7 d	152	0.2–2.2	0.2–2.2	108	0.2–2.2	0.2–2.2
8–14 d	180	0.3–3.0	0.3–3.0	124	0.1–2.9	0.1–2.9
15–30 d	390	0.2–3.5	0.2–3.5	290	0.2–5.0	0.2–5.0
31–60 d	74	0.3–2.7	0.3–2.7	184	0.2–2.1	0.2–2.1
61–180 d	75	0.5–1.9	0.5–1.9	102	0.6–1.9	0.6–1.9
0.5–<2 y	177	0.4–2.0	0.4–2.0	119	0.3–1.5	0.3–1.5
2–<6 y	189	0.3–1.2	0.3–1.2	134	0.5–1.1	0.5–1.1
6–<12 y	126	0.3–0.9	0.3–0.9	136	0.4–0.9	0.4–0.9
12–18 y	140	0.4–1.3	0.4–1.3	154	0.4–0.9	0.4–0.9

Specimen Type(s)	Whole blood (K_3/EDTA anticoagulant)
Reference(s)	Soldin SJ, Brugnara C, et al., eds. Pediatric reference ranges, 2nd ed. Washington, DC: AACC Press, 1997.
Method(s)	Coulter STKS (Beckman Coulter Inc., Brea, CA).
Comment(s)	Data from hospitalized patients (excluding hematology, oncology, and intensive care unit patients). A computerized approach adapted from the Hoffmann technique was used to remove outliers and obtain the 2.5th and 97.5th percentiles.

MONOCYTE COUNT (RELATIVE)

		Male			Female		
Test	Age	n	%	Number Fraction	n	%	Number Fraction
1	1–3 d	48	6.8–13.3	.068–.133	50	7.6–13.2	.076–.132
	4–7 d	33	8.2–13.1	.082–.131	29	5.9–12.0	.059–.120
	8–14 d	43	6.3–13.4	.063–.134	58	6.9–13.5	.069–.135
	15–30 d	137	7.1–13.7	.071–.137	98	6.0–15.9	.060–.159
	31–60 d	463	6.3–13.7	.063–.137	359	5.6–13.9	.056–.139
	61–180 d	1475	5.0–12.6	.050–.126	1166	4.8–12.4	.048–.124
	0.5–<2 y	5944	4.6–11.2	.046–.112	4614	4.4–10.6	.044–.106
	2–<6 y	6580	4.3–8.9	.043–.089	3419	4.1–8.7	.041–.087
	6–<12 y	5356	4.4–8.7	.044–.087	4729	4.0–8.0	.040–.080
	12–<18 y	3959	4.7–9.1	.047–.091	5121	4.1–8.0	.041–.080
	>18 y	1970	4.8–9.4	.048–.094	3060	4.2–7.9	.042–.079
2	1–3 d	172	0.8–11.9	.008–.119	114	1.4–16.3	.014–.163
	4–7 d	152	1.7–18.7	.017–.187	108	1.3–18.7	.013–.187
	8–14 d	180	1.9–20.8	.019–.208	124	2.3–22.2	.023–.222
	15–30 d	390	1.0–20.9	.010–.209	290	2.7–22.9	.027–.229
	31–60 d	74	4.9–19.6	.049–.196	184	2.4–19.1	.024–.191
	61–180 d	75	2.3–17.3	.023–.173	102	4.6–18.1	.046–.181
	0.5–<2 y	177	4.6–15.4	.046–.154	119	3.3–15.5	.033–.155
	2–<6 y	189	3.5–13.9	.035–.139	134	5.7–13.2	.057–.132
	6–<12 y	126	3.3–13.4	.033–.134	136	4.2–12.3	.042–.123
	12–18 y	140	3.1–12.5	.031–.125	154	3.8–11.2	.038–.112
3	1–14 d	38	9.1–13.9	.091–.139	28	7.1–12.0	.071–.120
	15–30 d	33	7.7–13.8	.077–.138	27	5.5–13.3	.055–.133
	31–60 d	147	5.9–13.9	.059–.139	109	6.0–12.0	.060–.120
	61–180 d	469	4.5–11.4	.045–.114	357	4.4–11.1	.044–.111
	0.5–<2 y	1558	4.1–10.2	.041–.102	1258	3.6–9.0	.036–.090
	2–<6 y	1637	3.5–7.8	.035–.078	1317	3.3–7.5	.033–.075
	6–<12 y	1491	3.7–7.9	.037–.079	1395	3.3–7.4	.033–.074
	12–<18 y	1415	3.8–8.2	.038–.082	1771	3.5–7.2	.035–.072
	>18 y	713	4.1–8.6	.041–.086	1077	3.4–7.0	.034–.070

Specimen Type(s)	1,2	Whole blood (K_3/EDTA anticoagulant)
	3	Whole blood (K_2/EDTA anticoagulant)
Reference(s)	1,2	Soldin SJ, Brugnara C, et al., eds. Pediatric reference ranges, 2nd ed. Washington, DC: AACC Press, 1997.
	3	Brugnara C. Unpublished data.
Method(s)	1	Bayer H*3 (Bayer Diagnostics, Tarrytown, NY).
	2	Coulter STKS (Beckman Coulter Inc., Brea, CA).
	3	Bayer ADVIA 120 (Bayer Diagnostics, Tarrytown, NY).
Comment(s)	1,3	Data from outpatients. A computerized approach adapted from the Hoffmann technique was used to remove outliers and obtain the 2.5th and 97.5th percentiles.
	2	Data from hospitalized patients (excluding hematology, oncology, and intensive care unit patients). A computerized approach adapted from the Hoffmann technique was used to remove outliers and obtain the 2.5th and 97.5th percentiles.

NEUTROPHIL COUNT (ABSOLUTE)

		Male			Female		
Test	**Age**	**n**	**× 10^3/μL**	**× 10^9/L**	**n**	**× 10^3/μL**	**× 10^9/L**
1	1–3 d	48	1.7–4.7	1.7–4.7	50	2.1–8.4	2.1–8.4
	4–7 d	32	1.9–4.1	1.9–4.1	29	1.8–5.1	1.8–5.1
	8–14 d	43	1.9–5.2	1.9–5.2	58	1.7–5.4	1.7–5.4
	15–30 d	137	1.5–3.6	1.5–3.6	98	1.3–4.3	1.3–4.3
	31–60 d	463	1.2–4.4	1.2–4.4	359	1.2–4.9	1.2–4.9
	61–180 d	1475	1.4–6.4	1.4–6.4	1166	1.4–6.7	1.4–6.7
	0.5–<2 y	5941	1.6–8.3	1.6–8.3	4607	1.8–9.1	1.8–9.1
	2–<6 y	6571	1.8–7.4	1.8–7.4	5674	1.8–6.8	1.8–6.8
	6–<12 y	5346	1.8–6.6	1.8–6.6	4721	1.8–6.7	1.8–6.7
	12–<18 y	3954	2.0–6.6	2.0–6.6	5110	2.3–6.9	2.3–6.9
	>18 y	1969	2.1–6.5	2.1–6.5	3055	2.5–7.0	2.5–7.0
2	1–3 d	173	2.7–9.0	2.7–9.0	114	3.4–11.3	3.4–11.3
	4–7 d	152	1.8–7.1	1.8–7.1	108	2.2–7.2	2.2–7.2
	8–14 d	180	3.2–11.4	3.2–11.4	124	1.7–7.6	1.7–7.6
	15–30 d	390	1.6–8.7	1.6–8.7	291	1.6–10.7	1.6–10.7
	31–60 d	302	1.7–6.0	1.7–6.0	184	1.5–5.8	1.5–5.8
	61–180 d	75	1.2–5.7	1.2–5.7	102	2.2–8.4	2.2–8.4
	0.5–<2 y	177	2.4–9.0	2.4–9.0	119	1.4–5.2	1.4–5.2
	2–<6 y	189	2.0–7.1	2.0–7.1	134	2.2–5.7	2.2–5.7
	6–<12 y	126	2.3–7.8	2.3–7.8	136	2.0–7.2	2.0–7.2
	12–18 y	140	2.8–11.1	2.8–11.1	154	2.3–6.7	2.3–6.7
3	1–14 d	38	2.1–7.5	2.1–7.5	27	2.2–4.1	2.2–4.1
	15–30 d	37	1.7–4.1	1.7–4.1	29	1.8–2.9	1.8–2.9
	31–60 d	148	1.1–3.2	1.1–3.2	111	1.3–3.8	1.3–3.8
	61–180 d	472	1.4–6.7	1.4–6.7	361	1.7–7.9	1.7–7.9
	0.5–<2 y	1569	1.8–9.4	1.8–9.4	1271	1.9–9.6	1.9–9.6
	2–<6 y	1651	2.1–7.9	2.1–7.9	1326	2.1–8.9	2.1–8.9
	6–<12 y	1503	2.1–7.3	2.1–7.3	1403	2.1–7.8	2.1–7.8
	12–<18 y	1429	2.1–7.0	2.1–7.0	1788	2.5–7.1	2.5–7.1
	>18 y	716	2.3–7.4	2.3–7.6	1092	2.5–7.5	2.5–7.5

Specimen Type(s)	1,2	Whole blood (K_3/EDTA anticoagulant)
	3	Whole blood (K_2/EDTA anticoagulant)
Reference(s)	1,2	Soldin SJ, Brugnara C, et al., eds. Pediatric reference ranges, 2nd ed. Washington, DC: AACC Press, 1997.
	3	Brugnara C. Unpublished data.
Method(s)	1	Bayer H*3 (Bayer Diagnostics, Tarrytown, NY).
	2	Coulter STKS (Beckman Coulter Inc., Brea, CA).
	3	Bayer ADVIA 120 (Bayer Diagnostics, Tarrytown, NY).
Comment(s)	1,3	Data from outpatients. A computerized approach adapted from the Hoffmann technique was used to remove outliers and obtain the 2.5th and 97.5th percentiles.
	2	Data from hospitalized patients (excluding hematology, oncology, and intensive care unit patients). A computerized approach adapted from the Hoffmann technique was used to remove outliers and obtain the 2.5th and 97.5th percentiles. To obtain the 2.5th and 97.5th percentiles, the statistical technique was modified to use the central 60% of the data for the Hoffmann plot.

NEUTROPHIL COUNT (RELATIVE)

		Male			Female		
Test	Age	n	%	Number Fraction	n	%	Number Fraction
1	1–3 d	48	24.1–50.3	.241–.503	50	23.1–58.4	.231–.584
	4–7 d	33	18.4–36.3	.184–.363	29	18.0–35.0	.180–.350
	8–14 d	43	18.3–36.3	.183–.363	58	17.1–33.1	.171–.331
	15–30 d	137	14.7–35.3	.147–.353	98	13.5–41.6	.135–.416
	31–60 d	462	14.2–40.0	.142–.400	359	13.6–44.5	.136–.445
	61–180 d	1475	16.3–51.6	.163–.516	1166	16.3–53.6	.163–.536
	0.5–<2 y	5946	21.3–66.7	.213–.667	4615	22.2–67.1	.222–.671
	2–<6 y	6580	30.3–74.3	.303–.743	5680	30.4–73.3	.304–.733
	6–<12 y	5356	36.3–74.3	.363–.743	4730	37.4–77.1	.374–.771
	12–<18 y	3959	41.2–75.5	.412–.755	5124	45.0–76.4	.450–.764
	>18 y	1970	38.9–75.1	.389–.751	3062	47.0–74.6	.470–.746
2	1–3 d	172	32.2–69.8	.322–.698	114	33.9–69.6	.339–.696
	4–7 d	152	25.6–67.7	.256–.677	108	21.6–59.0	.216–.590
	8–14 d	180	29.0–65.9	.290–.659	124	15.8–61.1	.158–.611
	15–30 d	390	18.7–58.8	.187–.588	291	17.3–61.0	.173–.610
	31–60 d	302	17.1–51.0	.171–.510	184	16.0–58.9	.160–.589
	61–180 d	75	11.9–53.4	.119–.534	102	21.8–61.2	.218–.612
	0.5–<2 y	177	24.6–70.9	.246–.709	119	20.7–61.8	.207–.618
	2–<6 y	189	34.3–78.6	.343–.786	134	32.3–69.7	.323–.697
	6–<12 y	126	40.3–72.9	.403–.729	136	39.5–76.9	.395–.769
	12–18 y	140	48.5–85.1	.485–.851	154	41.9–77.8	.419–.778
3	1–14 d	38	24.1–47.1	.241–.471	27	21.2–55.4	.212–.554
	15–30 d	37	18.4–32.4	.184–.324	29	17.0–40.9	.170–.409
	31–60 d	148	14.6–40.9	.146–.409	111	15.7–49.1	.157–.491
	61–180 d	472	17.0–55.5	.170–.555	361	18.6–60.0	.186–.600
	0.5–<2 y	1571	22.7–69.2	.227–.692	1272	23.8–69.3	.238–.693
	2–<6 y	1651	31.7–75.4	.317–.754	1328	33.6–77.5	.336–.775
	6–<12 y	1503	38.8–76.7	.388–.767	1406	38.7–76.7	.387–.767
	12–<18 y	1433	43.2–76.7	.432–.767	1786	46.4–75.6	.464–.756
	>18 y	716	47.2–77.6	.472–.776	1094	50.4–77.7	.504–.777

Specimen Type(s)	1,2	Whole blood (K_3/EDTA anticoagulant)
	3	Whole blood (K_2/EDTA anticoagulant)
Reference(s)	1,2	Soldin SJ, Brugnara C, et al., eds. Pediatric reference ranges, 2nd ed. Washington, DC: AACC Press, 1997.
	3	Brugnara C. Unpublished data.
Method(s)	1	Bayer H*3 (Bayer Diagnostics, Tarrytown, NY).
	2	Coulter STKS (Beckman Coulter Inc., Brea, CA).
	3	Bayer ADVIA 120 (Bayer Diagnostics, Tarrytown, NY).
Comment(s)	1,3	Data from outpatients. A computerized approach adapted from the Hoffmann technique was used to remove outliers and obtain the 2.5th and 97.5th percentiles.
	2	Data from hospitalized patients (excluding hematology, oncology, and intensive care unit patients). A computerized approach adapted from the Hoffmann technique was used to remove outliers and obtain the 2.5th and 97.5th percentiles. To obtain the 2.5th and 97.5th percentiles, the statistical technique was modified to use the central 60% of the data for the Hoffmann plot.

PLASMA FIBRINOGEN

		Male			Female		
Test	**Age**	**n**	**mg/dL**	**g/L**	**n**	**mg/dL**	**g/L**
1	1 d–18 y	131	150–302	1.50–3.02	96	184–366	1.84–3.66
2	1 d–18 y	204	119–346	1.19–3.46	146	116–293	1.16–2.93

Specimen Type(s)	1,2	Plasma (citrate anticoagulant)
Reference(s)	1,2	Soldin SJ, Brugnara C, et. al., eds. Pediatric reference ranges, 2nd ed. Washington, DC: AACC Press, 1997.
Method(s)	1	Reagents: Dade® Thromboplastin C+ (Dade International, Bedford, MA). Instrumentation: Electra 1000C (Hemoliance, Pleasantville, NY).
	2	Reagents: (Instrumentation Laboratories, Lexington, MA). Instrumentation: ACL 1000 (Instrumentation Laboratories, Lexington, MA).
Comment(s)	1	Data from outpatients. A computerized approach adapted from the Hoffmann technique was used to remove outliers and obtain the 2.5th and 97.5th percentiles.
	2	Data from hospitalized patients (excluding hematology, oncology, and intensive care unit patients). A computerized approach adapted from the Hoffmann technique was used to remove outliers and obtain the 2.5th and 97.5th percentiles.

PLATELET COUNT

		Male			Female		
Test	**Age**	**n**	**$\times 10^3/\mu L$**	**$\times 10^9/L$**	**n**	**$\times 10^3/\mu L$**	**$\times 10^9/L$**
1	1–3 d	55	164–351	164–351	60	234–346	234–346
	4–7 d	32	220–411	220–411	32	126–462	126–462
	8–14 d	48	226–587	226–587	59	265–557	265–557
	15–30 d	138	210–493	210–493	99	236–554	236–554
	31–60 d	442	275–567	275–567	339	295–615	295–615
	61–180 d	1393	275–566	275–566	1100	288–598	288–598
	0.5–<2 y	6248	219–452	219–452	4830	229–465	229–465
	2–<6 y	7576	204–405	204–405	6530	204–402	204–402
	6–<12 y	5450	194–364	194–364	4941	183–369	183–369
	12–<18 y	3935	165–332	165–332	5478	185–335	185–335
	>18 y	2048	143–320	143–320	4899	171–326	171–326
2	1–3 d	171	145[a]–262	145[a]–262	115	158[a]–300	158[a]–300
	4–7 d	151	128[a]–309	128[a]–309	107	108[a]–354	108[a]–354
	8–14 d	179	124[a]–367	124[a]–367	124	136[a]–411	136[a]–411
	15–30 d	386	153[a]–403	153[a]–403	288	122[a]–492	122[a]–492
	31–60 d	302	140[b]–557	140[b]–557	185	192[b]–574	192[b]–574
	61–180 d	75	270[b]–505[c]	270[b]–505[c]	102	135[b]–488[c]	135[b]–488[c]
	0.5–<2 y	177	200[b]–402[c]	200[b]–402[c]	119	238[b]–444[c]	238[b]–444[c]
	2–<6 y	190	197[b]–382[c]	197[b]–382[c]	135	213[b]–363[c]	213[b]–363[c]
	6–<12 y	126	175[b]–311[c]	175[b]–311[c]	135	130[b]–314[c]	130[b]–314[c]
	12–18 y	140	159–353	159–353	154	138–345	138–345
3	1–14 d	45	249–481	249–481	34	236–441	236–441
	15–30 d	45	253–493	253–493	36	315–506	315–506
	31–60 d	165	269–591	269–591	129	320–608	320–608
	61–180 d	524	307–619	307–619	387	304–574	304–574
	0.5–<2 y	2032	229–494	229–494	1684	238–497	238–497
	2–<6 y	2382	228–433	228–433	2022	229–435	229–435
	6–<12 y	1904	209–405	209–405	1779	204–406	204–406
	12–<18 y	1751	179–360	179–360	2410	201–363	201–363
	>18 y	1025	153–337	153–337	2496	177–354	177–354

4	0–14 d	56	218–419	218–419	46	144–449	144–449
	15–30 d	51	248–586	248–586	39	279–571	279–571
	31–60 d	115	229–562	229–562	72	331–597	331–597
	61–180 d	280	244–529	244–529	208	247–580	247–580
	0.5–<2 y	1208	206–445	206–445	1058	214–459	214–459
	2–<6 y	1218	202–403	202–403	1099	189–394	189–394
	6–<12 y	1289	206–369	206–369	1084	199–367	199–367
	12–<18 y	1481	175–332	175–332	1773	194–345	194–345
	>18 y	263	151–304	151–304	458	186–353	186–353

Specimen Type(s)	1,2,4	Whole blood (K_3/EDTA anticoagulant)
	3	Whole blood (K_2/EDTA anticoagulant)
Reference(s)	1,2	Soldin SJ, Brugnara C, et al., eds. Pediatric reference ranges, 2nd ed. Washington, DC: AACC Press, 1997.
	3	Brugnara C. Unpublished data.
	4	Children's National Medical Center, Washington, DC. Unpublished data.
Method(s)	1	Bayer H*3 (Bayer Diagnostics, Tarrytown, NY).
	2	Coulter STKS (Beckman Coulter Inc., Brea, CA).
	3	Bayer ADVIA 120 (Bayer Diagnostics, Tarrytown, NY).
	4	Sysmex XE-2100 (Sysmex Corporation, Kobe, Japan).
Comment(s)	1,3	Data from outpatients. A computerized approach adapted from the Hoffmann technique was used to remove outliers and obtain the 2.5th and 97.5th percentiles.
	2	Data from hospitalized patients (excluding hematology, oncology, and intensive care unit patients). A computerized approach adapted from the Hoffmann technique was used to remove outliers and obtain the 2.5th and 97.5th percentiles.
		[a]To obtain the 2.5th percentile, the statistical technique was modified to use the central 50% of the data for the Hoffmann plot.
		[b,c]To obtain the 2.5th and 97.5th percentiles, the statistical technique was modified to use the central 60% (instead of the usual 80%) of the data for the Hoffman plot.
	4	Performed on the XE-2100 on outpatients and ER patients. Hematology oncology clinic patients were excluded. Data were analyzed using SAS statistical software. Outliers were removed according to Chauvenet's Criterion (1). This criterion uses an exclusion factor based on the number of values in the sample set and is reapplied until no more outliers are excluded. The remaining data was truncated based on a method described by Hoffmann—the top and bottom 10% of data, based on an assumed normal distribution, was discarded (2). The method was modified to allow for computerized handling of the data. The 2.5th and 97.5th percentiles were calculated on the remaining 80% of the data.
		1. Young HD. Statistical treatment of experimental data. New York: McGraw-Hill, 1962:76–9, 162.
		2. Hoffmann RG. Statistics in the practice of medicine. JAMA 1963;185:864–73.

RED CELL COUNT

		Male			Female		
Test	Age	n	$\times 10^6/\mu L$	$\times 10^{12}/L$	n	$\times 10^6/\mu L$	$\times 10^{12}/L$
1	1–3 d	76	4.2–5.5	4.2–5.5	76	3.4–5.4	3.4–5.4
	4–7 d	49	3.9–5.4	3.9–5.4	39	3.5–5.5	3.5–5.5
	8–14 d	60	3.4–5.1	3.4–5.1	69	3.2–5.0	3.2–5.0
	15–30 d	168	3.1–4.6	3.1–4.6	116	3.1–4.6	3.1–4.6
	31–60 d	562	2.9–3.9	2.9–3.9	419	2.9–4.1	2.9–4.1
	61–180 d	1841	3.5–4.7	3.5–4.7	1363	3.4–4.6	3.4–4.6
	0.5–<2 y	8851	4.1–5.0	4.1–5.0	6934	4.1–4.9	4.1–4.9
	2–<6 y	11050	4.0–4.9	4.0–4.9	9752	4.0–4.9	4.0–4.9
	6–<12 y	7468	4.0–4.9	4.0–4.9	6804	4.0–4.9	4.0–4.9
	12–<18 y	5547	4.2–5.3	4.2–5.3	7977	4.0–4.9	4.0–4.9
	>18 y	3171	3.8–5.4	3.8–5.4	8606	3.8–4.8	3.8–4.8
2	1–3 d	174	3.3–5.0	3.3–5.0	115	3.3–5.3	3.3–5.3
	4–7 d	175	3.7–5.3	3.7–5.3	108	3.1–5.2	3.1–5.2
	8–14 d	182	3.5–5.2	3.5–5.2	124	3.6–4.9	3.6–4.9
	15–30 d	397	3.3–4.9	3.3–4.9	292	3.6–4.9	3.6–4.9
	31–60 d	74	3.1–4.1	3.1–4.1	185	3.1–4.7	3.1–4.7
	61–180 d	75	3.6–5.2	3.6–5.2	102	3.3–5.0	3.3–5.0
	0.5–<2 y	177	3.4–5.0	3.4–5.0	119	3.6–5.3	3.6–5.3
	2–<6 y	190	3.3[a]–4.8	3.3[a]–4.8	135	3.8[a]–4.9	3.8[a]–4.9
	6–<12 y	128	3.7[a]–4.9	3.7[a]–4.9	138	3.6[a]–4.9	3.6[a]–4.9
	12–18 y	140	3.3[a]–5.4	3.3[a]–5.4	154	3.4[a]–4.7	3.4[a]–4.7
3	1–14 d	48	3.8–5.3	3.8–5.3	34	3.9–5.2	3.9–5.2
	15–30 d	45	3.0–5.0	3.0–5.0	36	2.3–4.4	2.3–4.4
	31–60 d	172	2.9–3.8	2.9–3.8	132	3.0–4.1	3.0–4.1
	61–180 d	527	3.5–4.8	3.5–4.8	392	3.6–4.7	3.6–4.7
	0.5–<2 y	2056	4.1–6.0	4.1–6.0	1704	4.0–5.0	4.0–5.0
	2–<6 y	2403	4.1–5.0	4.1–5.0	2036	4.0–4.9	4.0–4.9
	6–<12 y	1931	3.9–5.0	3.9–5.0	1800	3.8–4.9	3.8–4.9
	12–<18 y	1731	3.8–5.3	3.8–5.3	2429	3.9–4.9	3.9–4.9
	>18 y	1042	3.4–5.4	3.4–5.4	2505	3.7–4.8	3.7–4.8

4	0–14 d	59	4.10–5.55	4.10–5.55	47	4.12–5.74	4.12–5.74
	15–30 d	53	3.16–4.63	3.16–4.63	39	3.32–4.80	3.32–4.80
	31–60 d	111	3.02–4.22	3.02–4.22	72	2.93–3.87	2.93–3.87
	61–180 d	282	3.43–4.80	3.43–4.80	214	3.45–4.75	3.45–4.75
	0.5–<2 y	1210	4.03–5.07	4.03–5.07	1049	3.97–5.01	3.97–5.01
	2–<6 y	1198	3.89–4.97	3.89–4.97	1087	3.84–4.92	3.84–4.92
	6–<12 y	1278	3.96–5.03	3.96–5.03	1084	3.90–4.96	3.90–4.96
	12–<18 y	1467	4.03–5.29	4.03–5.29	1762	3.93–4.90	3.93–4.90
	>18 y	264	4.18–5.48	4.18–5.48	467	3.70–4.87	3.70–4.87

Specimen Type(s)	1,2,4	Whole blood (K_3/EDTA anticoagulant)
	3	Whole blood (K_2/EDTA anticoagulant)
Reference(s)	1,2	Soldin SJ, Brugnara C, et al., eds. Pediatric reference ranges, 2nd ed. Washington, DC: AACC Press, 1997.
	3	Brugnara C. Unpublished data.
	4	Children's National Medical Center, Washington, DC. Unpublished data.
Method(s)	1	Bayer H*3 (Bayer Diagnostics, Tarrytown, NY).
	2	Coulter STKS (Beckman Coulter Inc., Brea, CA).
	3	Bayer ADVIA 120 (Bayer Diagnostics, Tarrytown, NY).
	4	Sysmex XE-2100 (Sysmex Corporation, Kobe, Japan).
Comment(s)	1,3	Data from outpatients. A computerized approach adapted from the Hoffmann technique was used to remove outliers and obtain the 2.5th and 97.5th percentiles.
	2	Data from hospitalized patients (excluding hematology, oncology, and intensive care unit patients). A computerized approach adapted from the Hoffmann technique was used to remove outliers and obtain the 2.5th and 97.5th percentiles.
		[a]To obtain the 2.5th percentile, the statistical technique was modified to use the central 60% of the data for the Hoffmann plot.
	4	Performed on the XE-2100 on outpatients and ER patients. Hematology oncology clinic patients were excluded. Data were analyzed using SAS statistical software. Outliers were removed according to Chauvenet's Criterion (1). This criterion uses an exclusion factor based on the number of values in the sample set and is reapplied until no more outliers are excluded. The remaining data was truncated based on a method described by Hoffmann—the top and bottom 10% of data, based on an assumed normal distribution, was discarded (2). The method was modified to allow for computerized handling of the data. The 2.5th and 97.5th percentiles were calculated on the remaining 80% of the data.
		1. Young HD. Statistical treatment of experimental data. New York: McGraw-Hill, 1962:76–9, 162.
		2. Hoffmann RG. Statistics in the practice of medicine. JAMA 1963;185:864–73.

RED CELL DISTRIBUTION WIDTH (RDW)

		Male		Female	
Test	Age	n	%	n	%
1	1–3 d	76	16.5–19.1	76	16.4–18.3
	4–7 d	48	15.9–17.9	39	15.7–18.5
	8–14 d	60	15.6–17.7	69	15.6–17.9
	15–30 d	168	15.2–17.4	116	15.1–17.9
	31–60 d	559	14.3–16.9	419	14.4–17.1
	61–180 d	1834	12.6–15.5	1360	12.4–15.1
	0.5–<2 y	8828	13.1–15.6	6905	13.0–15.3
	2–<6 y	11126	12.9–15.0	9724	12.7–14.7
	6–<12 y	7449	12.7–14.6	6787	12.5–14.4
	12–<18 y	5542	12.7–14.6	7985	12.5–14.5
	>18 y	3162	12.8–15.2	8592	12.7–15.1
2	1–3 d	174	14.9–18.5	115	14.7–17.5
	4–7 d	155	14.8–18.7	108	14.4–19.4
	8–14 d	182	14.6–19.2	124	14.4–19.9
	15–30 d	393	14.2–18.5	291	14.2–18.3
	31–60 d	74	12.9–17.0	185	14.0–17.5
	61–180 d	75	12.1–16.2	102	11.9–15.7
	0.5–<2 y	177	12.7–16.2	119	12.4–15.2
	2–<6 y	190	12.5–15.1	135	11.8–14.6
	6–<12 y	128	11.8–14.9	138	11.7–15.5
	12–18 y	140	12.1–18.4	154	12.0–17.4
3	1–14 d	48	15.2–18.0	34	15.3–16.6
	15–30 d	45	14.5–18.5	36	14.2–17.0
	31–60 d	169	13.8–16.3	132	14.0–16.4
	61–180 d	526	12.2–15.0	388	12.1–14.5
	0.5–<2 y	2055	12.7–15.0	1700	12.6–14.7
	2–<6 y	2396	12.5–14.3	2036	12.3–14.3
	6–<12 y	1922	12.2–14.2	1794	12.2–14.5
	12–<18 y	1752	12.3–14.7	2419	12.1–14.4
	>18 y	1040	12.4–15.0	2503	12.4–15.3

4	0–14 d	66	14.8–17.0	49	14.6–17.3
	15–30 d	56	14.3–16.8	43	14.4–16.2
	31–60 d	111	13.8–16.1	75	13.6–15.8
	61–180 d	298	12.4–15.3	223	12.2–14.3
	0.5–<2 y	1212	12.9–15.6	1064	12.7–15.1
	2–<6 y	1179	12.5–14.9	1074	12.4–14.9
	6–<12 y	1243	12.3–14.1	1090	12.2–14.4
	12–<18 y	1423	12.4–14.5	1737	12.3–14.6
	>18 y	257	12.3–14.3	155	12.4–15.1

Specimen Type(s)	1,2,4	Whole blood (K_3/EDTA anticoagulant)
	3	Whole blood (K_2/EDTA anticoagulant)
Reference(s)	1,2	Soldin SJ, Brugnara C, et al., eds. Pediatric reference ranges, 2nd ed. Washington, DC: AACC Press, 1997.
	3	Brugnara C. Unpublished data.
	4	Children's National Medical Center, Washington, DC. Unpublished data.
Method(s)	1	Bayer H*3 (Bayer Diagnostics, Tarrytown, NY).
	2	Coulter STKS (Beckman Coulter Inc., Brea, CA).
	3	Bayer ADVIA 120 (Bayer Diagnostics, Tarrytown, NY).
	4	Sysmex XE-2100 (Sysmex Corporation, Kobe, Japan).
Comment(s)	1,3	Data from outpatients. A computerized approach adapted from the Hoffmann technique was used to remove outliers and obtain the 2.5th and 97.5th percentiles.
	2	Data from hospitalized patients (excluding hematology, oncology, and intensive care unit patients). A computerized approach adapted from the Hoffmann technique was used to remove outliers and obtain the 2.5th and 97.5th percentiles.
	4	Performed on the XE-2100 on outpatients and ER patients. Hematology oncology clinic patients were excluded. Data were analyzed using SAS statistical software. Outliers were removed according to Chauvenet's Criterion (1). This criterion uses an exclusion factor based on the number of values in the sample set and is reapplied until no more outliers are excluded. The remaining data was truncated based on a method described by Hoffmann—the top and bottom 10% of data, based on an assumed normal distribution, was discarded (2). The method was modified to allow for computerized handling of the data. The 2.5th and 97.5th percentiles were calculated on the remaining 80% of the data. 1. Young HD. Statistical treatment of experimental data. New York: McGraw-Hill, 1962:76–9, 162. 2. Hoffmann RG. Statistics in the practice of medicine. JAMA 1963;185:864–73.

RED CELL DISTRIBUTION WIDTH, SD (RDW)

	Male		Female	
Age	n	fL	n	fL
0–14 d	57	51.0–61.7	47	51.4–65.7
15–30 d	47	46.3–57.3	42	47.2–59.8
31–60 d	110	43.9–52.8	75	43.0–55.0
61–180 d	273	35.3–45.7	210	35.2–45.1
0.5–<2 y	1179	35.3–42.8	1030	34.9–42.4
2–<6 y	1146	35.1–41.7	1042	34.9–42.0
6–<12 y	1224	35.1–41.7	1036	35.5–41.8
12–<18 y	1391	36.7–43.8	1704	37.1–44.2
>18 y	255	37.8–46.1	449	38.4–47.7

Specimen Type(s)	Whole blood (K_3/EDTA anticoagulant)
Reference(s)	Children's National Medical Center, Washington, DC. Unpublished data.
Method(s)	Sysmex XE-2100 (Sysmex Corporation, Kobe, Japan).
Comment(s)	Performed on the XE-2100 on outpatients and ER patients. Hematology oncology clinic patients were excluded. Data were analyzed using SAS statistical software. Outliers were removed according to Chauvenet's Criterion (1). This criterion uses an exclusion factor based on the number of values in the sample set and is reapplied until no more outliers are excluded. The remaining data was truncated based on a method described by Hoffmann—the top and bottom 10% of data, based on an assumed normal distribution, was discarded (2). The method was modified to allow for computerized handling of the data. The 2.5th and 97.5th percentiles were calculated on the remaining 80% of the data. 1. Young HD. Statistical treatment of experimental data. New York: McGraw-Hill, 1962:76–9, 162. 2. Hoffmann RG. Statistics in the practice of medicine. JAMA 1963;185:864–73.

RETICULOCYTE CELLULAR HEMOGLOBIN CONTENT (CHR)

	Male		Female	
Age	**n**	**pg/cell**	**n**	**pg/cell**
1 d–<2 y	132	22.5–31.8	104	23.9–30.9
2–<6 y	127	25.1–32.0	92	26.4–32.1
6–<12 y	133	23.6–33.9	116	25.1–33.3
12–<18 y	211	27.0–33.2	221	28.2–33.9
>18 y	214	30.1–34.6	402	27.1–35.2

Specimen Type(s)	Whole blood (K_2/EDTA anticoagulant)
Reference(s)	Brugnara C. Unpublished data.
Method(s)	Bayer ADVIA 120 (Bayer Diagnostics, Tarrytown, NY).
Comment(s)	Data from outpatients. A computerized approach adapted from the Hoffmann technique was used to remove outliers and obtain the 2.5th and 97.5th percentiles.

RETICULOCYTE COUNT (ABSOLUTE)

	Male			Female		
Age	**n**	**$\times 10^3/\mu L$**	**$\times 10^9/L$**	**n**	**$\times 10^3/\mu L$**	**$\times 10^9/L$**
1–<6 mo	66	37–104	37–104	24	52–120	52–120
6 mo–<2 y	68	29–89	29–89	78	35–92	35–92
2–<6 y	126	29–80	29–80	92	43–83	43–83
6–<12 y	129	39–106	39–106	115	37–93	37–93
12–<18 y	204	39–100	39–100	216	40–102	40–102
>18 y	213	43–85	43–85	392	46–102	46–102

Specimen Type(s)	Whole blood (K_2/EDTA anticoagulant)
Reference(s)	Brugnara C. Unpublished data.
Method(s)	Bayer ADVIA 120 (Bayer Diagnostics, Tarrytown, NY).
Comment(s)	Data from outpatients. A computerized approach adapted from the Hoffmann technique was used to remove outliers and obtain the 2.5th and 97.5th percentiles.

RETICULOCYTE COUNT (RELATIVE)

	Male			Female		
Age	n	%	Number Fraction	n	%	Number Fraction
1–3 d	22	2.2–4.8	.022–.048	26	2.1–3.7	.021–.037
4–30 d	58	0.4–2.7	.004–.027	29	0.4–2.0	.004–.020
31–60 d	67	0.9–3.8	.009–.038	34	1.5–3.2	.015–.032
61–180 d	138	0.9–3.1	.009–.031	98	1.1–2.9	.011–.029
0.5–<2 y	602	0.8–2.0	.008–.020	439	0.9–2.0	.009–.020
2–<6 y	966	0.8–2.0	.008–.020	770	0.8–2.1	.008–.021
6–<12 y	688	0.7–2.2	.007–.022	570	0.8–2.8	.008–.028
12–18 y	480	0.8–2.2	.008–.022	558	0.8–2.2	.008–.022

Specimen Type(s)	Whole blood (K_3/EDTA anticoagulant)
Reference(s)	Soldin SJ, Brugnara C, et al., eds. Pediatric reference ranges, 2nd ed. Washington, DC: AACC Press, 1997.
Method(s)	Bayer H*3 (Bayer Diagnostics, Tarrytown, NY).
Comment(s)	Data from outpatients. A computerized approach adapted from the Hoffmann technique was used to remove outliers and obtain the 2.5th and 97.5th percentiles.

WHITE CELL COUNT

		Male			Female		
Test	Age	n	× 10^3/µL	× 10^9/L	n	× 10^3/µL	× 10^9/L
1	1–3 d	76	6.8–13.3	6.8–13.3	76	8.0–14.3	8.0–14.3
	4–7 d	48	8.3–14.1	8.3–14.1	39	8.8–14.8	8.8–14.8
	8–14 d	59	8.2–14.4	8.2–14.4	68	8.4–15.4	8.4–15.4
	15–30 d	167	7.4–14.6	7.4–14.6	115	8.3–14.7	8.3–14.7
	31–60 d	560	6.7–14.2	6.7–14.2	415	7.0–15.1	7.0–15.1
	61–180 d	1832	6.9–15.7	6.9–15.7	1359	6.8–16.0	6.8–16.0
	0.5–<2 y	8751	6.2–14.5	6.2–14.5	6829	6.4–15.0	6.4–15.0
	2–<6 y	11056	5.3–11.5	5.3–11.5	9675	5.3–11.5	5.3–11.5
	6–<12 y	7431	4.5–10.5	4.5–10.5	6718	4.7–10.3	4.7–10.3
	12–<18 y	5519	4.5–10.0	4.5–10.0	7971	4.8–10.1	4.8–10.1
	>18 y	3133	4.4–10.2	4.4–10.2	8571	4.9–10.0	4.9–10.0
2	1–3 d	174	5.5–15.0	5.5–15.0	115	8.4–16.6	8.4–16.6
	4–7 d	175	6.2–13.0	6.2–13.0	108	7.9–16.0	7.9–16.0
	8–14 d	182	7.8[a]–19.3[b]	9.0–20.7	124	7.3[a]–16.5[b]	9.2–17.8
	15–30 d	369	7.2[a]–19.9[b]	9.0–22.1	292	7.2[a]–18.4[b]	9.1–19.6
	31–60 d	392	7.4–14.3	7.4–14.3	185	8.2–16.3	8.2–16.3
	61–180 d	75	8.3–15.9	8.3–15.9	102	7.9–13.4	7.9–13.4
	0.5–<2 y	177	7.8–14.8	7.8–14.8	119	6.3–12.6	6.3–12.6
	2–<6 y	190	5.0–13.2	5.0–13.2	135	6.4–11.1	6.4–11.1
	6–<12 y	128	5.0–11.8	5.0–11.8	138	4.8–10.7	4.8–10.7
	12–<18 y	140	5.1–15.5	5.1–15.5	154	4.9–9.1	4.9–9.1
3	1–14 d	47	8.6–14.9	8.6–14.9	34	8.3–17.6	8.3–17.6
	15–30 d	45	7.4–13.3	7.4–13.3	36	6.9–15.0	6.9–15.0
	31–60 d	169	6.0–13.7	6.0–13.7	131	6.1–13.8	6.1–13.8
	61–180 d	526	6.6–15.6	6.6–15.6	391	6.8–16.2	6.8–16.2
	0.5–<2 y	2042	6.3–15.4	6.3–15.4	1694	6.4–15.5	6.4–15.5
	2–<6 y	2392	5.5–12.3	5.5–12.3	2024	5.5–12.5	5.5–12.5
	6–<12 y	1911	4.7–11.3	4.7–11.3	1791	4.9–11.4	4.9–11.4
	12–<18 y	1756	4.4–10.5	4.4–10.5	2416	4.9–10.4	4.9–10.4
	>18 y	1027	4.3–10.4	4.3–10.4	2500	4.9–10.7	4.9–10.7

4	0–14 d	56	8.04–15.40	8.04–15.40	44	8.16–14.56	8.16–14.56
	15–30 d	53	7.80–15.91	7.8–15.91	35	8.36–14.42	8.36–14.42
	31–60 d	111	8.14–14.99	8.14–14.99	75	7.05–14.68	7.05–14.68
	61–180 d*	67	6.51–13.32	6.51–13.32	44	6.00–13.25	6.00–13.25
	0.5–<2 y*	537	5.98–13.51	5.98–13.51	466	6.48–13.02	6.48–13.02
	2–<6 y	1194	5.14–13.38	5.14–13.38	1075	4.86–13.18	4.86–13.18
	6–<12 y	1276	4.31–11.00	4.31–11.00	1068	4.27–11.40	4.27–11.40
	12–<18 y	1454	3.84–9.84	3.84–9.84	1736	4.19–9.43	4.19–9.43
	>18 y	261	3.91–8.77	3.91–8.77	458	4.37–9.68	4.37–9.68

Specimen Type(s)	1,2,4	Whole blood (K_3/EDTA anticoagulant)
	3	Whole blood (K_2/EDTA anticoagulant)
Reference(s)	1,2	Soldin SJ, Brugnara C, et al., eds. Pediatric reference ranges, 2nd ed. Washington, DC: AACC Press, 1997.
	3	Brugnara C. Unpublished data.
	4	Children's National Medical Center, Washington, DC. Unpublished data.
Method(s)	1	Bayer H*3 (Bayer Diagnostics, Tarrytown, NY).
	2	Coulter STKS (Beckman Coulter Inc., Brea, CA).
	3	Bayer ADVIA 120 (Bayer Diagnostics, Tarrytown, NY).
	4	Sysmex XE-2100 (Sysmex Corporation, Kobe, Japan).
Comment(s)	1,3	Data from outpatients. A computerized approach adapted from the Hoffmann technique was used to remove outliers and obtain the 2.5th and 97.5th percentiles.
	2	Data from hospitalized patients (excluding hematology, oncology, and intensive care unit patients). A computerized approach adapted from the Hoffmann technique was used to remove outliers and obtain the 2.5th and 97.5th percentiles. To obtain the 2.5th percentile, the statistical technique was modified to use the central 60% of the data for the Hoffmann plot. [a]Central 80% used for 2.5th percentile. [b]Central 50% used for 97.5th percentile.
	4	*Excludes ER specimens. Performed on the XE-2100 on outpatients and ER patients. Hematology oncology clinic patients were excluded. Data were analyzed using SAS statistical software. Outliers were removed according to Chauvenet's Criterion (1). This criterion uses an exclusion factor based on the number of values in the sample set and is reapplied until no more outliers are excluded. The remaining data was truncated based on a method described by Hoffmann—the top and bottom 10% of data, based on an assumed normal distribution, was discarded (2). The method was modified to allow for computerized handling of the data. The 2.5th and 97.5th percentiles were calculated on the remaining 80% of the data. 1. Young HD. Statistical treatment of experimental data. New York: McGraw-Hill, 1962:76–9, 162. 2. Hoffmann RG. Statistics in the practice of medicine. JAMA 1963;185:864–73.